M58

Internal Corrosion Control in Water Distribution Systems

Second Edition

American Water Works Association

Manual of Water Supply Practices—M58, Second Edition

Internal Corrosion Control in Water Distribution Systems

Project Manager: Melissa Valentine
Cover Art: Melanie Yamamoto
Production: Glacier Publishing Services, Inc.
Manuals Specialist: Sue Bach

Library of Congress Cataloging-in-Publication Data has been applied for

Names: Giani, Richard E., author. | Hill, Christopher P. Internal corrosion control in water distribution systems. | American Water Works Association.
Title: M58--internal corrosion control in water distribution systems / by Richard E. Giani.
Other titles: Internal corrosion control in water distribution systems
Description: Second edition. | Denver, CO : American Water Works Association, [2017] | Earlier edition: Internal corrosion control in water distribution systems / [by Christopher P. Hill, Abigail F. Cantor]. c2011. | Includes bibliographical references and index.
Identifiers: LCCN 2017002003 | ISBN 9781625762023
Subjects: LCSH: Water-pipes--Corrosion. | Drinking water--Contamination. | Water--Distribution. | Corrosion and anti-corrosives--Testing.
Classification: LCC TD491 .G49 2017 | DDC 628.1/44--dc23 LC record available at https://lccn.loc.gov/2017002003

ISBN-13 978-1-62576-202-3
eISBN-13 978-1-61300-404-3

Printed in the United States of America
American Water Works Association
6666 West Quincy Avenue
Denver, CO 80235-3098
awwa.org

Contents

Figures

This page intentionally blank.

Tables

This page intentionally blank.

Preface

This is the second edition of AWWA Manual M58, *Internal Corrosion Control in Water Distribution Systems*, prepared by the Distribution System Water Quality Committee. It provides practical guidance to aid public water systems in designing, implementing, and maintaining an effective corrosion control program. This manual is intended to help readers understand the factors that influence corrosion, assess corrosion-related impacts, and develop a strategy to implement and maintain effective corrosion control in the water distribution system.

The increased focus surrounding corrosion-related water quality and infrastructure impacts in water distribution systems has generated new research and information in this area. As such, this update of AWWA Manual M58 is a reflection of those advancements, including changes in regulatory requirements, the addition of pertinent case studies and examples, and incorporation of other new material reflective of changes in the industry. In the future, the AWWA Distribution System Water Quality Committee will continue to update this manual as new issues and questions arise.

Editor's note: Throughout this manual, references are made to Awwa Research Foundation (AwwaRF). In 2009, the foundation changed its name to Water Research Foundation. Any publication prior to 2009 will reflect the foundation's original name.

This page intentionally blank.

Acknowledgments

This second edition of AWWA Manual M58 was written through the persistent, dedicated work of the following authors:

J. Boxall, University of Sheffield, Sheffield, U.K.
R. Brown, Cornwell Engineering Group, Long Beach, Calif.
A. Cantor, Process Research Solutions, LLC, Madison, Wis.
M. Friedman, Confluence Engineering Group, LLC, Seattle, Wash.
R. Giani, CH2M, Lees Summit, Mo.
A. Hill, Confluence Engineering Group, LLC, Seattle, Wash.
C. Hill, Arcadis, Tampa, Fla.
S. Husband, University of Sheffield, Sheffield, U.K.
A. Jacque, Water Quality Investigations, LLC, Mount Horeb, Wis.
B. Lakin, McMillen Jacobs Associates, New York, N.Y.
M. McFadden, M3P Services, LLC, Denver, Colo.
S. Randtke, University of Kansas, Lawrence, Kans.
R. Slabaugh, Arcadis, Indianapolis, Ind.
V. Speight, University of Sheffield, Sheffield, U.K.
E. Turner, North Texas Municipal Water District, Wylie, Texas

The authors would like to acknowledge the support of the following organizations in preparing this manual:

American Water Works Association, Denver, Colo.
Arcadis, Indianapolis, Ind., and Tampa, Fla.
CH2M, Lees Summit, Mo.
Confluence Engineering Group, LLC, Seattle, Wash.
Process Research Solutions, LLC, Madison, Wis.

The authors would also like to acknowledge the following individuals who provided editorial and technical comments:

J. Cuppett, Water Research Foundation, Denver, Colo.
L. Furatian, University of British Columbia, Vancouver, B.C.
M. Horton, United States Pipe and Foundry Company, Bessemer, Ala.
P. Kwan, HDR Engineering, Inc., Seattle, Wash.
M. Larsen, Kennedy/Jenks Consultants, Federal Way, Wash.
I. Zhu, Evoqua Water Technologies, Warrendale, Pa.

The following individuals provided peer review of this manual. Their knowledge and efforts are gratefully appreciated:

F. Lemieux, Health Canada, Ottawa, Ont.
Q. Muylwyk, CH2M, Toronto, Ont.
M. Roberts, Hazen and Sawyer, Greensboro, N.C.
M. Rodgers, Cleveland Division of Water, Cleveland, Ohio
K. Smith, City of Houston, Houston, Texas
J. Swertfeger, Greater Cincinnati Water Works, Cincinnati, Ohio

Chapter 1

Overview of Internal Corrosion Impacts in Drinking Water Distribution Systems

Christopher P. Hill, Arcadis

INTRODUCTION

Prior to developing a corrosion control program, it is important to first gain an understanding of the factors that influence internal corrosion and metal release in drinking water distribution systems. Afterward, implementation of an effective corrosion control program can be accomplished in eight steps (Table 1-1).

Selection of an effective corrosion control strategy is not a one-time event. Changes in source water quality, treatment, distribution, operational, or regulatory changes may require a system to reevaluate, revise, optimize, or restore a corrosion control program. To assist the reader in navigating the process described in Table 1-1, this manual is organized accordingly.

PURPOSE OF THIS MANUAL

Internal corrosion of drinking water distribution and home plumbing systems has long been a recognized issue facing the drinking water industry—one that has many health, water quality, and economic implications.

Table 1-1 Eight steps to implementing an effective corrosion control program

Step	Discussed in Chapter(s)
1. Develop an understanding of the general concepts behind internal corrosion and metal release in drinking water distribution systems	1, 2
2. Determine the extent and magnitude of corrosion	3
3. Determine the possible causes of corrosion	2, 3
4. Assess corrosion control alternatives	4
5. Develop a corrosion control strategy	4, 5
6. Implement a corrosion control program	3, 6
7. Monitor the effectiveness of the corrosion control program	3, 6
8. Optimize the corrosion control program	5, 6

This manual was written with the intent to provide a practical overview of internal corrosion issues, to identify appropriate corrosion response and control methods, and to develop corrosion control monitoring programs. It is not intended to provide detailed summaries of corrosion chemistry. References to additional resources are given where the reader might find more detailed information useful.

Although this manual briefly touches on internal corrosion of nonmetallic pipe surfaces, such as asbestos–cement and cement mortar–lined ductile- or cast-iron pipe, it primarily focuses on corrosion of metal pipe surfaces, solders, and plumbing fixtures, such as those composed of lead, copper, and iron. For the purposes of this manual, the term *corrosion* refers not only to the electrochemical phenomenon that causes metal loss from pipe surfaces but also to the dissolution of existing pipe scales and corrosion by-products. Internal corrosion—that is, corrosion on the interior surface of metal pipes and fixtures—is the focus of this manual because of the potential for metal release to adversely impact distributed water quality.

CUSTOMER AND INFRASTRUCTURE IMPACTS OF INTERNAL CORROSION

Internal corrosion can cause degradation of water quality, infrastructure performance and structural failures, and scaling and reequilibration issues. It may have substantial economic impacts on water utilities and consumers alike. This section discusses the most common issues resulting from internal corrosion of drinking water distribution and home plumbing systems.

Water Quality Deterioration

Internal corrosion of distribution system piping and home plumbing may cause several water quality problems, including potential health concerns, discoloration, and taste and odor issues. These problems are primarily the result of corrosion of metal pipe surfaces, pipe solder, and plumbing fixtures or dissolution of existing pipe scales. However, some problems may be attributable to corrosion of nonmetallic system components (e.g., increases in asbestos concentrations).

Health concerns. The majority of the health concerns associated with internal corrosion are related to the release of trace metal concentrations (lead, copper, cadmium, etc.) from corroding metal surfaces (ATSDR 2007). However, corrosion can affect the ability to maintain a disinfectant residual in the distribution system, which may result in environments suitable for biofilm growth that may harvest pathogens, such as *Legionella* and various mycobacteria. The potential health concerns related to increased metal concentrations

Courtesy of DC Water

Figure 1-1 Red-water sample from a US distribution system

in drinking water are discussed further in the "Regulatory Impacts" section later in this chapter.

Color. Many of the color issues typically encountered in drinking water are attributable to internal corrosion (Kirmeyer et al. 2000a, 2000b). Corrosion of cast-iron pipe or dissolution of existing scale on cast-iron pipe may result in rust-colored water or red water as a result of the presence of ferric iron (Fe(III)). Red water can stain laundry and plumbing fixtures.

Red water (Figure 1-1) may occur as a result of "hydraulic entrainment" (i.e., the suspension and transport of loose corrosion deposits resulting from changes in flow or direction of flow) or "iron uptake" (i.e., the release of ferrous ions at the pipe surface and subsequent oxidation to ferric iron and precipitation in the bulk water) (Smith et al. 1998). Alternating periods of stagnation, anaerobic conditions, and temperature have been observed to cause red water, as have significant changes in alkalinity and chlorine residual (oxidation–reduction potential [ORP]). Maintaining consistent water quality and maintaining conditions that produce a hard iron scale are critical to minimizing the potential for red water (Smith et al. 1998, Reiber 2006).

Iron corrosion may also result in yellow or black water, which is a consequence of the presence of ferrous iron (Fe(II)). Ferrous iron may occur in waters with low dissolved oxygen concentrations (Kirmeyer et al. 2000a, 2000b). Black water may also be the result of the presence of manganese (Mn) in source waters or contribution from chemicals added during treatment, such as iron-based coagulants.

Copper corrosion may result in blue or green water, which can also stain bathroom fixtures and hair. Blue water may be a consequence of the presence of either dissolved or particulate copper corrosion by-products. Dissolved copper by-products are primarily the result of low pH and can usually be eliminated by raising the pH to above 7.5 (Edwards et al. 2000). Hard, high-alkalinity groundwaters are particularly aggressive toward copper materials (Schock and Lytle 2011). Edwards et al. (1999) found that high-alkalinity waters (>74 mg/L as calcium carbonate) produced the highest 90th-percentile copper levels at pH <7.8.

However, it is often more difficult to determine the cause of particulate copper corrosion by-products, which are found in many home plumbing systems. Particulate copper corrosion by-products are thought to be primarily the result of microbiological, chemical, or physical factors, or some combination of factors that frequently result in pitting corrosion and/or pipe failure in addition to blue water (Edwards et al. 2000, Bremer et al. 2001).

Gray or black water may also be attributable to hot water heaters. Corrosion of zinc in galvanized piping may result in a milky appearance (Kirmeyer et al. 2000a, 2000b).

Taste and odor. Dissolved and particulate iron may cause taste and odor issues at low concentrations. The presence of iron has resulted in metallic taste (Khiari et al. 2002). In addition to metallic taste, cast-iron corrosion may also produce musty tastes and odors (Kirmeyer et al. 2000a, 2000b). Aluminum and zinc may contribute to an astringent mouthfeel, and zinc may also result in a sour taste. Reactions of disinfectants with cement–mortar linings in ductile-iron pipe may also result in astringent, oily (rancid), pine, and phenolic odors (Khiari et al. 2002). High pH may cause the release of phenols from corroding asbestos–cement pipe, which can react with chlorine to form chlorophenols (Kirmeyer et al. 2000a, 2000b).

Infrastructure Impacts

In addition to affecting water quality, internal corrosion may also impact distribution and plumbing infrastructure. Most notably, internal corrosion may result in failures of home plumbing systems, causing extensive property damage. Deterioration of distribution system piping and valves as well as other in-line devices may also result from corrosion.

Home plumbing failures. The most significant home plumbing failures that occur because of internal corrosion are copper pitting and pinhole leaks. Pitting corrosion, though not well understood, not only damages copper plumbing, but the associated leakage may also cause substantial damage to homes and result in mold growth and other issues that may pose health concerns (Edwards 2004). Table 1-2 summarizes the conditions under which potential copper corrosion has been traditionally thought to occur.

In addition to pitting corrosion, microbiologically influenced corrosion (MIC) of copper piping may also cause pitting and failure of home plumbing (Cantor et al. 2003, 2006) and MIC is often overlooked as a contributor to corrosion of plumbing systems. MIC is more likely to occur in plumbing systems with long stagnation times, in sulfide-containing waters, and in areas in which there is little to no disinfectant residual remaining (Jacobs et al. 1998, Bremer et al. 2001, Cantor et al. 2003).

Distribution system failures. Corrosion of distribution system piping may lead to pipe and valve failures resulting in increased water losses. In areas of low or negative pressure, inflow and infiltration are possible, as are bacterial and other contamination from surrounding soils and groundwaters. Failure of corroded valves may also cause operational issues, such as the inability to isolate water mains in the event of a line break, loss of pressure, reduced ability to conduct distribution flushing, and creation of stagnant sections or areas of low flow in the distribution system (e.g., if a valve stem breaks in the closed position). Failures of meters and other in-line devices may reduce system revenues because of inaccurate meter readings and may minimize the ability to accurately determine water loss or identify main breaks.

Scaling and Reequilibration

Scaling and reequilibration are two key factors that may impact corrosion and system performance, reequilibration being the transition period following a change in distribution system water quality and lasting until equilibrium is re-established. Formation of protective metal scales is the primary corrosion control mechanism regardless of the technology employed. For example, when pH and alkalinity adjustment are used as the primary lead control strategy, the objective is to form metal solids (e.g., lead carbonate and lead oxides)

Table 1-2 Summary of potential copper corrosion issues, causes, and solutions

	Type of Corrosion			
Characteristic of Corrosion	Uniform Corrosion	Type I Pitting (Cold Water)	Type II Pitting (Hot Water)	Type III Pitting (Soft Water)
Pit shape	No pits	Deep and narrow	Narrower than Type I	Wide and shallow
Type of problem present with corrosion	Blue or green water, high by-product release	Pipe failure	Pipe failure	Blue water, voluminous by-product releases, pipe blockage
Scale morphology on attacked surface	Tarnished copper surface or loose powdery scale	Underlying Cu_2O with overlying malachite, calcite, or other basic copper salts, occasionally CuCl underlies Cu_2O	Underlying Cu_2O with overlying bronchantite ($Cu_4SO_4(OH)_6$), some malachite ($Cu_2CO_3(OH)_2$)	Underlying Cu_2O with overlying bronchantite ($Cu_4SO_4(OH)_6$), some malachite ($Cu_2CO_3(OH)_2$)
Water quality	Low pH (<7.2) or waters with high alkalinity (>250 mg/L)	Hard, cold well waters between pH 7.0 and 7.8, high sulfate relative to chlorides and bicarbonate, high CO_2	Hot waters, pH <7.2, high sulfate relative to bicarbonate, occasional Mn deposits	Soft waters, pH >8.0
Initiating factors	None noted	Stagnation early in pipe life, deposits within pipe such as dirt or carbon fines, high free chlorine residuals, water softeners, residual aluminum carryover from coagulation process	Higher temperatures, high free chlorine residuals, residual aluminum carryover from coagulation process	Stagnation early in pipe life, pH >8.0, residual aluminum carryover from coagulation process, low free chlorine residual
Mitigating factors and treatments	Reduce alkalinity and raise pH or use orthophosphate	Increase bicarbonate and raise pH	Raise pH and/ or increase bicarbonate	Avoid stagnation early in pipe life, increase hardness and alkalinity, elevate chlorine residual to >0.5 mg/L

Source: *Adapted from Edwards et al. 1994*

on the pipe surface and thereby prevent corrosion, or dissolution, of lead. Similarly, the objective of orthophosphate addition is the formation of a lead phosphate layer (Vik et al. 1996).

Controlling the type of scale formed and subsequently maintaining the stability of those scales are key to an effective corrosion control program. Iron scales, for example, are present in either the ferrous (Fe^{2+}) form or ferric (Fe^{3+}) form. Ferrous iron scales are much softer than ferric iron scales and are more likely to contribute to red water (Sarin et al. 2000, Reiber 2006). Creating conditions in which harder ferric iron scales are formed and maintained is ideal to maintaining water quality.

After a protective scale is formed in distribution piping or home plumbing, maintaining scale stability is essential to effective corrosion control treatment. Changes in distribution system water quality may result in reequilibration of existing scales causing red water, increased tap lead or copper concentrations, or other issues.

For example, in 2000, the Washington Aqueduct (Washington, D.C.) switched from free chlorine to chloramine to reduce disinfection by-product concentrations in its customer agencies' systems. As a result of this adjustment, the ORP of the finished water changed, causing a shift in the speciation of the existing lead scales and elevated lead concentrations in the District of Columbia Water (DC Water) service area (USEPA 2007). The conversion to chloramine was just one factor that contributed to the lead release. Many utilities operating chloraminated systems have not experienced similar results.

Economic Issues

Internal corrosion may have numerous economic or cost impacts. Excess corrosion can result in tuberculation or excess calcium precipitation, reducing hydraulic capacity, increasing pumping costs, and reducing fire flow capabilities. Failure of distribution system piping and home plumbing may necessitate costly repairs, not only of failed piping but also of other assets damaged because of the pipe failure.

There are also more subtle economic issues associated with corrosion. Increased lead and copper levels may not only result in negative publicity but may also have the potential to bring about litigation. The costs associated with legal action may be substantial. Water quality issues resulting from corrosion (e.g., color, taste and odor, and metals release) may also cause negative customer perceptions and not only affect (reduce) usage but also cost utilities potential future customers.

REGULATORY IMPACTS

Lead and Copper

In the United States, the Lead and Copper Rule (LCR) was promulgated in 1991 with the purpose of reducing drinking water exposures to lead and copper (56 FR 26460; *Federal Register* 1991a). The purpose of the LCR is to protect public health by minimizing lead and copper levels in drinking water, primarily by reducing water corrosivity (USEPA 2004a). Lead has been demonstrated to cause delays in physical and mental development in infants and children and has been linked to deficits in attention span and learning abilities, even at low levels (USEPA 2006a, ATSDR 2007). Short-term copper exposure may cause gastrointestinal distress; long-term exposure may cause liver or kidney damage. In individuals with Wilson's disease, which causes the body to retain copper, copper may cause severe brain damage, liver failure, and death (ATSDR 2004, NIH 2006).

The US Environmental Protection Agency (USEPA) estimates that approximately 20 percent of human lead exposure is from lead in drinking water (USEPA 2006a). The LCR established a maximum contaminant level goal (MCLG) of zero for lead and an MCLG of 1.3 mg/L for copper. The rule established a National Primary Drinking Water Regulation for lead and copper that consisted of a treatment technique requirement including corrosion control treatment, source water treatment, lead service line (LSL) replacement, and public education. The rule set an action level (AL) of 0.015 mg/L for lead and 1.3 mg/L for copper. If the 90th percentile concentration for lead or copper is above the AL, public water systems may be required to initiate water quality parameter monitoring for key corrosion control parameters, install corrosion control treatment, begin source water monitoring or treatment, replace the LSL, or undertake a public education program. USEPA issued guidance to assist utilities in complying with the corrosion control treatment requirements of the LCR in 1992 (USEPA 1992) and provided updated guidance to assist in identification and maintenance of optimal corrosion control treatment in 2016 (USEPA 2016).

Internal corrosion of LSLs, brass meters and plumbing fixtures, and copper plumbing contributes to the concentrations of lead and copper in drinking water. Pure lead pipe (LSLs), lead solder, and brass with greater than 8 percent lead were banned by the 1986 Safe Drinking

Water Act Amendments (Dudi et al. 2005). "Lead-free" brass, however, was allowed to contain as much as 2 to 8 percent lead by weight under this amendment. Greenville Utilities in Greenville, N.C., exceeded the lead AL in 2004 and 2005 even though the system does not contain any LSLs. In this case, the exceedance was attributed to leaching from lead-based solder and plumbing fixtures containing lead (Landers 2006). California banned the use of pipe solder with lead content greater than 0.2 percent in 1987. In 2010, California law defined "lead free" as less than 0.25 percent weighted average and required that all pipes, fittings, and fixtures meet this definition (California DTSC 2009). Similarly, effective January 4, 2014, the federal Reduction of Lead in Drinking Water Act redefined lead-free as containing no more than 0.25% lead as a weighted average on the wetted surface and required "any pipe or plumbing fitting or fixture, any solder, or any flux" used in the installation or repair of any public water system, or any plumbing in a residential or nonresidential facility providing water for human consumption, to be lead free (USEPA 2013a).

In Canada, the low-lead requirements are integrated into plumbing standards that are referenced in the National Plumbing Code of Canada (NPC). In November 2013, a modification of the NPC was approved whereby plumbing standards that required fittings meet the 0.25% weighted average lead limit were added to the 2010 version of NPC as an interim change. The interim change was published in late 2013 (NRCC 2013) and subsequently in the 2015 version of the NPC (NRCC 2015).

Impact of the LCR on tap-water lead concentrations. A review conducted by USEPA of LCR compliance monitoring data for public water systems serving more than 3,300 people shows that since 2000, fewer than 4 percent of those systems have exceeded the lead AL (USEPA 2004b). Prior to 2002, states were not required to report 90th percentile lead concentrations to USEPA unless those values exceeded the AL. Consequently, it is difficult to compare the tap-water lead concentrations in the years following implementation of the LCR to current levels. However, USEPA did evaluate data from 166 large public water systems that exceeded the AL following initial LCR monitoring in 1992 and 1993 (USEPA 2006d). Based on the results of subsequent monitoring conducted by these same utilities between 2000 and 2004, only 15 of those systems continued to exceed the AL (Figure 1-2).

A closer look at the data presented in Figure 1-2 reveals that tap-water lead concentrations have decreased significantly in those systems that initially exceeded the lead AL. Figure 1-3 shows that, for those 166 large systems that exceeded the AL in 1992–1993, the average 90th percentile lead concentration across all 166 systems surveyed decreased from nearly 32 μg/L to 8.2 μg/L. Over that same period, the maximum 90th percentile lead concentration decreased from 211 μg/L to 84 μg/L.

Revisions to the Lead and Copper Rule. USEPA revised the LCR in 2007 (72 FR 57781; *Federal Register* 2007) with the intent of enhancing implementation in the areas of monitoring, treatment, customer awareness, and LSL replacement. In addition, the 2007 revisions intended to improve public education by providing drinking water customers with "meaningful, timely, and useful information needed to reduce exposure to lead in drinking water."

The most significant revisions to the LCR provided clarification regarding sample collection and reporting. Specifically, the LCR revisions required that all compliance monitoring results, including those above the required number of samples, be included in the 90th percentile determination and that all samples be taken within the same calendar year for those systems on reduced monitoring. The remainder of the revisions addressed customer notification for those residents in a utility's sampling program and provisions to address sampling sites with known LSLs that were previously "tested out" due to sample results below the AL.

Long-term revisions to the Lead and Copper Rule. In the years since LCR implementation, a variety of noteworthy research within the drinking water community has provided new insights in lead and copper corrosion and has led to concerns that the

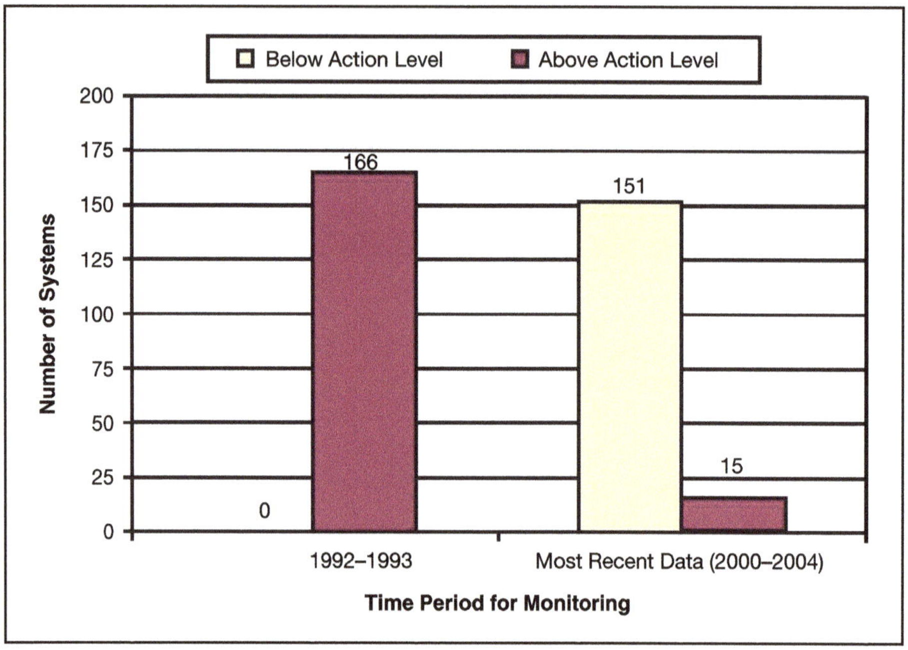

Source: *USEPA 2006d*

Figure 1-2 Number of large US water systems exceeding the lead action level

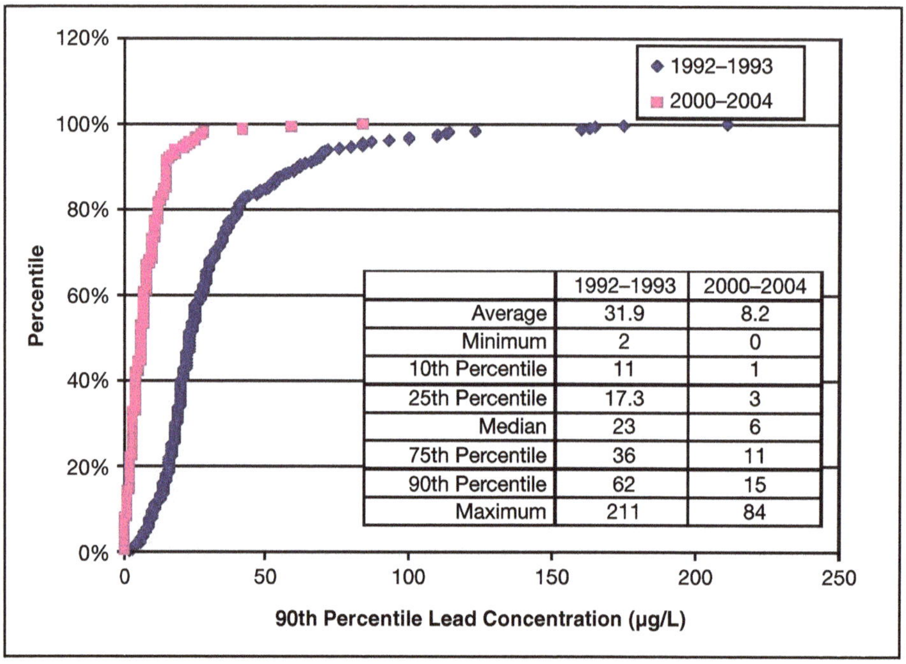

	1992–1993	2000–2004
Average	31.9	8.2
Minimum	2	0
10th Percentile	11	1
25th Percentile	17.3	3
Median	23	6
75th Percentile	36	11
90th Percentile	62	15
Maximum	211	84

Figure 1-3 Comparison of 90th percentile LCR monitoring data for 166 large public water systems

current LCR may systematically underestimate lead and copper levels in drinking water (e.g., Del Toral et al. 2013). The Lead and Copper Rule Working Group (LCRWG) of the National Drinking Water Advisory Council (NDWAC) was formed to provide recommendations to USEPA regarding the Long-Term Revisions to the Lead and Copper Rule. The LCRWG was tasked with addressing the following issues:

- Sample site selection criteria
- Lead sampling protocols
- Public education for copper
- Measures to ensure optimal corrosion control treatment
- LSL replacement

The LCRWG met between March 2014 and June 2015 and issued its recommendations to USEPA in August 2015 (NDWAC 2015). Among its findings, the *Report of the Lead and Copper Rule Working Group to the National Drinking Water Advisory Council* concluded that the anticipated Long-Term Revisions to the Lead and Copper Rule represent a significant opportunity for "removing sources of lead in contact with drinking water and for reducing exposure to lead from drinking water." Further, the LCRWG noted that eliminating the sources of lead in contact with drinking water would require creative funding solutions, as total replacement is expected to cost billions of dollars nationwide. It also concluded that more robust public outreach and education is necessary.

In making its recommendations to USEPA,

> The LCRWG took the following considerations, among others, into account in making recommendations for revisions to the current LCR.
>
> There is no safe level of lead. Lead can pose health risks to anyone, but there are heightened risks for pregnant women, infants and young children and other vulnerable populations with both acute and chronic exposures. Effective elimination of leaded materials in contact with water and minimization of exposure to lead in drinking water is a shared responsibility; public water systems (PWSs), consumers, building owners, public health officials and others each have important roles to play. The lack of resources to reduce the sources of exposure in some communities, however, also raises important questions of disparate impact and environmental justice. Thus, creative financing mechanisms will be needed.
>
> The LCR should remain a treatment technique rule, but it can be improved based on the scientific knowledge that has emerged since the current LCR was promulgated. Corrosion control treatment is complicated, and will vary based on specific circumstances in each public water system. Thus, regular updates to guidance by EPA based on the latest science and the creation of a national clearinghouse of information both for the public and for PWSs are needed.
>
> The LCRWG considered but did not quantify the cost implications of its recommendations. An important factor in the group's deliberations was the principle that PWS and state resources should be focused on actions that achieve the greatest public health protection. Recognizing that lead service line (LSL) replacement programs will be costly in some locations, the LCRWG also encouraged PWSs to incorporate anticipated costs into their capital improvement program as appropriate to their situation, and urges states to include the costs of LSL replacement in their criteria for allocation of Drinking Water State Revolving Funds.
>
> The LCRWG specifically recommends that EPA revise the LCR to:
>
> - Require proactive lead service line replacement programs, which set replacement goals, effectively engage customers in implementing those

goals, and provide improved access to information about LSLs, in place of current requirements in which LSLs must be replaced only after a lead action level (AL) exceedance;

- Establish more robust public education requirements for lead and LSLs, by updating the Consumer Confidence Report (CCR), adding targeted outreach to consumers with lead service lines and other vulnerable populations (pregnant women and families with infants and young children), and increasing the information available to the public;
- Strengthen corrosion control treatment (CCT), retaining the current rule requirements to re-assess CCT if changes to source water or treatment are planned, adding a requirement to review updates to EPA guidance to determine if new scientific information warrants changes;
- Modify monitoring requirements to provide for consumer requested tap samples for lead and to utilize results of tap samples for lead to inform consumer action to reduce the risks in their homes, to inform the appropriate public health agency when results are above a designated household action level, and to assess the effectiveness of CCT and/or other reasons for elevated lead results;
- Tailor water quality parameters (WQPs) to the specific CCT plan for each system, and increase the frequency of WQP monitoring for process control;
- Establish a health-based, household action level that triggers a report to the consumer and to the applicable health agency for follow up;
- Separate the requirements for copper from those for lead and focus new requirements where water is corrosive to copper; and
- Establish appropriate compliance and enforcement mechanisms. (NDWAC 2015)

USEPA is currently considering the LCRWG recommendations and is anticipated to issue a proposed rule in 2017. While not obligated to adopt the working group recommendations, USEPA is likely to include the suggestions, or some form of them, in the revised rule, most notably the recommendations for proactive LSL replacement, enhanced WQP monitoring, and more robust public education.

Health Canada guidelines. The safety of drinking water supplies in Canada is a responsibility that is shared between the provincial/territorial, federal, and municipal governments. The day-to-day responsibility of providing safe drinking water to the public generally rests with the provinces and territories while the municipalities usually oversee the day-to-day operations of the treatment facilities. Health Canada develops its *Guidelines for Canadian Drinking Water Quality* in partnership with the provinces and territories. These guidelines are used by every jurisdiction in Canada and are the basis for establishing provincial drinking water quality requirements for all Canadians (Health Canada 2015).

In Canada, the guideline for lead is a maximum acceptable concentration (MAC) of 0.01 mg/L (10 μg/L) based on a provisional tolerable weekly intake of lead established by the World Health Organization (Health Canada 1992). However, Health Canada is reviewing the lead guideline to include the most recent health information as well as sampling and monitoring strategies that address exposure from drinking water (Health Canada 2013). Treatment-based guidance for corrosion control was published in 2009 (Health Canada 2009). The recommended sampling protocol found in this guidance uses stagnation times to capture the highest levels of lead to provide the basis for a system-wide assessment of the corrosion control efficacy.

Iron and Manganese

USEPA established secondary maximum contaminant levels (SMCLs) of 0.3 mg/L for iron and 0.05 mg/L for manganese, which represent a "reasonable goal for drinking water quality" (*Federal Register* 1979, 1991b; USEPA 2013b). SMCLs control contaminants that primarily affect the aesthetic quality of drinking water and are not federally enforceable, although state primacy agencies have the authority to include them in state drinking water regulations. When present at sufficiently high levels, iron may result in rusty color, deposit of sediment, metallic taste, and reddish or orange staining. Similarly, Health Canada's established aesthetic objective for iron in drinking water is ≤0.3 mg/L (≤300 μg/L). A MAC has not been set for iron (Health Canada 1978).

Much of the iron present in water is because of natural mineral deposits, and when present in source waters at concentrations above the SMCL, iron is frequently removed at the water treatment facility. If not removed, source-water iron may result in red water in the distribution system. However, internal corrosion of unlined cast-iron pipe, galvanized pipe, or dissolution of existing pipe scales in iron pipe may also result in aesthetic issues in drinking water distribution systems.

Iron, manganese, and aluminum scales in water distribution systems have adsorptive properties and can contain substantially higher concentrations of trace metals, such as lead, copper, cadmium, and zinc, than are present in the bulk water entering the system (Friedman at al. 2009, Deshommes et al. 2010, McFadden et al. 2011, Camara et al. 2013, Schock et al. 2014). Release of these metals can occur because of changes in pH and ORP associated with source water blending or source changeovers, treatment process changes, disinfectant residual loss, or biological reactions. These bulk water chemistry changes can cause desorption of contaminants from iron and manganese scales or can result in chemical destabilization of existing scales (Friedman at al. 2009). Physical releases of existing iron scales may trigger red-water events, which may contain elevated concentrations of trace metals. Physical release can be caused by changes in distribution system hydraulics, such as sudden changes in demand, flushing activities, fire flow events, main breaks, or other changes in flow patterns.

Cadmium

Short-term exposure to high concentrations of cadmium may result in nausea, vomiting, diarrhea, muscle cramps, salivation, sensory disturbances, liver injury, convulsions, shock, and renal failure. Long-term exposures may cause kidney, liver, bone, and blood damage (USEPA 2006b). Consequently, USEPA has established an MCLG of 0.005 mg/L for cadmium as well as a maximum contaminant level (MCL) of 0.005 mg/L. A MAC of 0.005 mg/L (5 μg/L) for cadmium in drinking water has been established on the basis of health considerations (Health Canada 1986).

Cadmium is primarily present in drinking waters as a result of erosion of natural deposits but may also be present because of leaching from cement–mortar-lined pipes (Guo et al. 1998, Berend and Trouwborst 1999) or from corrosion of galvanized piping. It has also been observed that the amount of lead and cadmium released from galvanized piping decreases with the age of the pipe as a consequence of the formation of a passivating layer on the pipe surface (Meyer 1980).

Zinc

Zinc in drinking water is primarily an aesthetic concern because it may result in a metallic taste. Consequently, USEPA has established an SMCL of 5 mg/L for zinc. An aesthetic objective has been established for zinc in drinking water of ≤5.0 mg/L (Health Canada 1987).

Zinc may be present in drinking water due to erosion of natural deposits or from dezincification of brass plumbing fixtures. Several studies of low-alkalinity waters have found that iron and zinc were the principal corrosion by-products of galvanized plumbing (Dangel 1975, JMM 1982).

The zinc coating on galvanized pipe may contain lead, copper, cadmium, chromium, aluminum, barium, and other impurities (AwwaRF and TZW 1996). There are a number of standards regarding galvanized pipe, most notably by ASTM International in the United States. As a result of these impurities, corrosion of galvanized pipe may result in the release of trace metal concentrations.

Asbestos

Asbestos in drinking water is most commonly present as a result of degradation of asbestos–cement piping used in drinking-water distribution systems. Though it may be present in natural soils, it does not often migrate to groundwater through soils (USEPA 2006c). Short-term exposure to asbestos in drinking water is not known to cause any health issues; long-term exposure may result in an increased risk of developing benign intestinal polyps. USEPA has established an MCLG of 7 million fibers per liter (MFL) and an MCL of 7 MFL for asbestos.

Biological Regrowth and Chlorine Demand

In the United States, the Surface Water Treatment Rule was finalized in 1989 and requires maintenance of a disinfectant residual in the distribution system, among other requirements (54 FR 27486; *Federal Register* 1989a). The Total Coliform Rule (TCR) imposed monitoring requirements for total coliform as well as for fecal coliform and *Escherichia coli* (54 FR 27544; *Federal Register* 1989b). The Revised Total Coliform Rule further established an MCL for *E. coli* and uses *E. coli* and total coliforms to initiate a "find and fix" approach to address fecal contamination that could enter into the distribution system (78 FR 10269; *Federal Register* 2013). It requires public water systems to perform assessments to identify sanitary defects and subsequently take action to correct them. Biological regrowth and chlorine demand may be affected by corrosion in many ways and vice versa. Chlorine and other oxidants impact the ORP, which is a factor in maintaining effective corrosion control. Conversely, corrosion by-products can exert an oxidant demand, diminishing distribution system disinfectant residuals and resulting in increased potential for microbiological regrowth. Microbiological regrowth not only may threaten compliance with the TCR but may also result in increased potential for MIC to occur.

Release of Trace Metals From Cementitious Coatings

Barium, cadmium, chromium, and aluminum have been found to leach from cement–mortar lining in distribution piping (Guo et al. 1998, Berend and Trouwborst 1999). While this leaching represents a potential water quality and public health concern, it is not the focus of this manual.

ADDITIONAL READING

The reader is advised to become more knowledgeable in the following areas:

- The water quality and operational factors that influence internal corrosion and metal release in drinking water distribution systems

- The potential water quality, health-related infrastructure, and regulatory concerns associated with internal corrosion
- The secondary impacts associated with changes in source-water quality, treatment, or distribution operations and the potential for those changes to affect existing corrosion control effectiveness (i.e., reequilibration of existing pipe scales)

References for this chapter give detailed information on these topics and the other topics discussed. The reader is encouraged to follow up on the referenced material that is publicly available to obtain keener insight into this discussion.

REFERENCES

ATSDR (Agency for Toxic Substances and Disease Registry). 2004. *Toxicological Profile for Copper.* Atlanta, Ga.: ATSDR, Public Health Service, US Department of Health and Human Services. Available at: www.atsdr.cdc.gov/toxprofiles/tp132.pdf

ATSDR. 2007. *Toxicological Profile for Lead.* Atlanta, Ga.: ATSDR, Public Health Service, US Department of Health and Human Services. Available at: www.atsdr.cdc.gov/toxprofiles/tp13.pdf

AwwaRF and TZW (Awwa Research Foundation and DVGW-Technologiezentrum Wasser). 1996. *Internal Corrosion of Water Distribution Systems*, 2nd ed. Denver, Colo.: AwwaRF.

Berend, K., and T. Trouwborst. 1999. Cement–mortar pipes as a source of aluminum. *Journal AWWA*, 91(7):91–100.

Bremer, P.J., B.J. Webster, and D.B. Wells. 2001. Biocorrosion of copper in potable water. *Journal AWWA*, 93(8):82–91.

California DTSC (Department of Toxic Substance Control). 2009. Requirements for Low Lead Plumbing Products in California. Available at: https://dtsc.ca.gov/PollutionPrevention/upload/lead-in-plumbing-fact-sheet.pdf

Camara, E., K.R. Montreuil, A.K. Knowles, and G.A. Gagnon. 2013. Role of the water main in lead service line replacement: A utility case study. *Journal AWWA*, 105(8):E423–E431.

Cantor, A.F., J.B. Bushman, and M.S. Glodoski. 2003. A new awareness of copper pipe failures in water distribution systems. In *Proceedings of the AWWA Water Quality Technology Conference*. Denver, Colo.: AWWA.

Cantor, A.F., J.B. Bushman, M.S. Glodoski, E. Kiefer, R. Bersch, and H. Wallenkamp. 2006. Copper pipe failure by microbiologically influenced corrosion. *Materials Performance*, 45(6):38.

Dangel, R.A. 1975. *Study of Corrosion Products in the Seattle Water Department Distribution System*. Report 670/2-75-036. Cincinnati, Ohio: USEPA.

Del Toral, M.A., A. Porter, and M.R. Schock. 2013. Detection and evaluation of elevated lead release from service lines: A field study. *Environmental Science and Technology*, 47:9300.

Deshommes, E., L. Laroche, S. Nour, C. Cartier, and M. Prévost. 2010. Source and occurrence of particulate lead in tap water. *Water Research*, 44(12):3734–3744.

Dudi, A., M. Schock, N. Murray, and M. Edwards. 2005. Lead leaching from inline brass devices: A critical evaluation of the existing standard. *Journal AWWA*, 97(8):66–78.

Edwards, M. 2004. Written testimony to Chairman Tom Davis and US House of Representatives. Committee on Government Reform, May 5.

Edwards, M., J.F. Ferguson, and S.H. Reiber. 1994. The pitting corrosion of copper. *Journal AWWA*, 86(7):74–90.

Edwards, M., S. Jacobs, and D.M. Dodrill. 1999. Desktop guidance for mitigating Pb and Cu corrosion by-products. *Journal AWWA*, 91(5):66–77.

Edwards, M., S. Jacobs, and R.J. Taylor. 2000. The blue water phenomenon. *Journal AWWA*, 92(7):72–82.

Federal Register. 1979. National Secondary Drinking Water Regulations; Final Rule. *Federal Register*, 44(140):42195.

Federal Register. 1989a. National Primary Drinking Water Regulations; Filtration, Disinfection; Turbidity, Giardia lamblia, Viruses, Legionella, and Heterotrophic Bacteria, Final Rule. *Federal Register*, 54(124):27486–27567.

Federal Register. 1989b. National Primary Drinking Water Regulations; Total Coliforms (Including Fecal Coliform and *E. coli*); Final Rule. *Federal Register*, 54(124):27544–27568.

Federal Register. 1991a. National Primary Drinking Water Regulations; Lead and Copper; Final Rule. *Federal Register*, 56(110):26460.

Federal Register. 1991b. National Secondary Drinking Water Regulations; Final Rule. *Federal Register*, 56(20):3526.

Federal Register. 2007. National Primary Drinking Water Regulations for Lead and Copper: Short-Term Regulatory Revisions and Clarifications. *Federal Register*, 72(195):57781–57820.

Federal Register. 2013. National Primary Drinking Water Regulations: Revisions to the Total Coliform Rule; Final Rule. *Federal Register*, 78(30):10270–10365.

Friedman, M.J., A.S. Hill, S.H. Reiber, R.L. Valentine, and G.V. Korshin. 2009. *Assessment of Inorganics Accumulation in Drinking Water System Scales and Sediments*. Project 3118. Denver, Colo.: Water Research Foundation.

Guo, Q., P.J. Toomuluri, and J.O. Eckert Jr. 1998. Leachability of regulated metals from cement–mortar linings. *Journal AWWA*, 90(3):62–73.

Health Canada. 1978. Iron. In *Guidelines for Canadian Drinking Water Quality* (supporting documentation). Ottawa, Ont.: Health Canada, Bureau of Chemical Hazards, Environmental Health Directorate.

Health Canada. 1986. Cadmium. In *Guidelines for Canadian Drinking Water Quality* (supporting documentation). Ottawa, Ont.: Health Canada, Bureau of Chemical Hazards, Environmental Health Directorate.

Health Canada. 1987. Zinc. In *Guidelines for Canadian Drinking Water Quality* (supporting documentation). Ottawa, Ont.: Health Canada, Bureau of Chemical Hazards, Environmental Health Directorate.

Health Canada. 1992. *Guidelines for Canadian Drinking Water Quality: Guideline Technical Document—Copper*. Ottawa, Ont.: Health Canada, Water Quality and Health Bureau, Healthy Environments and Consumer Safety Branch.

Health Canada. 2009. *Guidance on Controlling Corrosion in Drinking Water Distribution Systems*. Ottawa, Ont.: Health Canada, Water, Air and Climate Change Bureau, Healthy Environments and Consumer Safety Branch.

Health Canada. 2013. *Risk Management Strategy for Lead*. Ottawa, Ont.: Health Canada. Available at: www.hc-sc.gc.ca/ewh-semt/pubs/contaminants/prms_lead-psgr_plomb/index-eng.php

Health Canada. 2015. *Guidelines for Canadian Drinking Water Quality*—Summary Table. Available at: www.hc-sc.gc.ca/ewh-semt/pubs/water-eau/sum_guide-res_recom/index-eng.php#t

Jacobs, S., S. Reiber, and M. Edwards. 1998. Sulfide-induced copper corrosion. *Journal AWWA*, 90(7):62–73.

JMM (J.M. Montgomery Consulting Engineering Inc.). 1982. *Internal Corrosion Mitigation Study*. Final Report. Prepared for the Portland, Ore., Water Bureau.

Khiari, D., S. Barrett, R. Chinn, A. Bruchet, P. Piriou, L. Matia, F. Ventura, I. Suffet, T. Gittelman, and P. Leutweiler. 2002. *Distribution Generated Taste and Odor Phenomena*. Denver, Colo.: AwwaRF and AWWA.

Kirmeyer, G.J., M. Friedman, and J. Clement. 2000a. *Guidance Manual for Maintaining Distribution System Water Quality*. Denver, Colo.: AwwaRF and AWWA.

Kirmeyer, G.J., G. Pierson, J. Clement, A. Sandvig, V. Snoeyink, W. Kriven, and A. Camper. 2000b. *Distribution System Water Quality Changes Following Corrosion Control Strategies*. Denver, Colo.: AwwaRF and AWWA.

Landers, J. 2006. Getting the lead out. *Public Works*, 137(4):30–32.

McFadden, M., R. Giani, P. Kwan, and S.H. Reiber. 2011. Contributions to drinking water lead from galvanized iron corrosion scales. *Journal AWWA*, 103(4):76.

Meyer, E. 1980. Beeinträchtigung der Trinkwassergüte durch Anlagenteile der Hausinstallation—Bestimmung des Schwermetalleintrags in das Trinkwasser durch Korrosionsvorgänge in metallischen Rohren. *DVGW-Schriftenreihe Wasser Nr.*, 23:113–131.

NDWAC (National Drinking Water Advisory Council). 2015. *Report of the Lead and Copper Rule Working Group to the National Drinking Water Advisory Council*. Final, National Drinking Water Advisory Council. Available at: www.awwa.org/Portals/0/files/legreg/documents/NDWACRecommendations.pdf

NIH (National Institutes of Health). 2006. *Wilson Disease*. National Digestive Diseases Information Clearinghouse, National Institute of Diabetes and Digestive and Kidney Diseases, National Institutes of Health. Available at: http://digestive.niddk.nih.gov/ddiseases/pubs/wilson/index.htm

NRCC (National Research Council Canada). 2013. *National Plumbing Code of Canada*. Ottawa, Ont.: National Research Council Canada.

NRCC. 2015. *National Plumbing Code of Canada*. Ottawa, Ont.: National Research Council Canada.

Reiber, S. 2006. Corrosion, red water, and distribution system water quality. In *Proceedings of the AWWA Annual Conference and Exposition*. Denver, Colo.: AWWA.

Sarin, P., J. Bebee, M.A. Becket, K.K. Jim, D.A. Lytle, J.A. Clement, W.M. Kriven, and V.L. Snoeyink. 2000. Mechanism of release of iron from corroded iron/steel pipes in water distribution systems. In *Proceedings of the AWWA Annual Conference and Exposition*. Denver, Colo.: AWWA.

Schock, M., and D. Lytle. 2011. Internal corrosion and deposition control. In *Water Quality and Treatment: A Handbook on Drinking Water*, 6th ed. Edited by J.K. Edzwald. Denver, Colo.: McGraw-Hill and American Water Works Association.

Schock, M.R., A. Cantor, S. Triantafyllidou, M.K. Desantis, and K.G. Scheckel. 2014. Importance of pipe deposits to Lead and Copper Rule compliance. *Journal AWWA*, 106(7):E336–E349.

Smith, S.E., T. Ta, D.M. Holt, A. Delanoue, and J. Colbourne. 1998. Minimizing red water in drinking water distribution systems. In *Proceedings of the AWWA Water Quality Technology Conference*. Denver, Colo.: AWWA.

USEPA (US Environmental Protection Agency). 1992. *Lead and Copper Rule Guidance Manual, Volume II: Corrosion Control Treatment*. EPA 811-B-92-002. Washington, D.C.: USEPA, Office of Water.

USEPA. 2004a. *Lead and Copper Rule: Quick Reference Guide*. EPA 816-F-04-009. Washington, D.C.: USEPA, Office of Water.

USEPA. 2004b. Summary of Lead Action Level Exceedances for Medium (3,300–50,000) and Large (>50,000) Public Water Systems. Available at: www.epa.gov/safewater/lcrmr/pdfs/data_lcmr_leadsummary_0522041.pdf

USEPA. 2006a. Lead in Drinking Water: Basic Information. Available at: www.epa.gov/safewater/lead/basicinformation.html

USEPA. 2006b. Consumer Fact Sheet: Cadmium. Available at: www.epa.gov/OGWDW/contaminants/dw_contamfs/cadmium.html

USEPA. 2006c. Consumer Fact Sheet: Asbestos. Available at: www.epa.gov/OGWDW/contaminants/dw_contamfs/asbestos.html

USEPA. 2006d. Lead 90th Percentile Levels for 166 Large Water Utilities—Then and Now. Available at: www.epa.gov/safewater/lcrmr/reductionplan_comparrison.html

USEPA. 2007. *Elevated Lead in DC Drinking Water.* EPA 815-R-07-021. Washington, D.C.: USEPA, Office of Water.

USEPA. 2013a. Reduction of Lead in Drinking Water Act. Available at: https://www.epa.gov/dwstandardsregulations/section-1417-safe-drinking-water-act-prohibition-use-lead-pipes-solder-and

USEPA. 2013b. Secondary Drinking Water Regulations: Guidance for Nuisance Chemicals. Available at: http://water.epa.gov/drink/contaminants/secondarystandards.cfm

USEPA. 2016. *Optimal Corrosion Control Treatment Evaluation Technical Recommendations for Primacy Agencies and Public Water Systems.* EPA 816-B-16-003. Washington, D.C.: USEPA, Office of Water.

Vik, E.A., R.A. Ryder, I. Wagner, and J.F. Ferguson. 1996. Mitigation of corrosion effects. In *Internal Corrosion of Water Distribution Systems,* 2nd ed. Denver, Colo.: AwwaRF and AWWA.

This page intentionally blank.

Chapter 2

Fundamentals of Internal Corrosion and Metal Release

Abigail F. Cantor, Process Research Solutions, LLC
Melinda Friedman, Confluence Engineering Group, LLC
Andrew Jacque, Water Quality Investigations, LLC
Rebecca Slabaugh, Arcadis
Vanessa Speight, Stewart Husband, and *Joby Boxall*, University of Sheffield

INTRODUCTION

There are many misunderstandings about corrosion in water distribution systems. Just using the word *corrosion* gives an incomplete picture of what is occurring. The concern is actually the transfer of metal from pipe walls into the bulk water that results in elevated metal concentrations in drinking water. Corrosion, especially uniform corrosion that is typically the focus of corrosion control studies, is only one of several mechanisms that can transfer metal from pipes (Table 2-1).

This chapter describes the common mechanisms that may cause metal release into water. Understanding and managing these mechanisms are the keys to achieving what is commonly called *corrosion control*.

Table 2-1 Chapter 2 key points

Uniform corrosion, which is typically the focus of corrosion control studies, is only one of many mechanisms that can cause the release of lead, copper, and other metals into water.
Uniform Corrosion
• In uniform corrosion, the anodic and cathodic sites move around dynamically on the pipe wall, resulting in uniform loss of metal. • Calcium carbonate is inadequate as a protective barrier against uniform corrosion. • Naturally occurring constituents such as dissolved inorganic carbon, hydroxides, and oxygen can form compounds with dissolved metals. Those compounds create protective barriers against uniform corrosion to varying degrees, as they are mostly influenced by pH and alkalinity of the water. The adequacy of the pipe films as protective barriers depends on the solubility of the compounds formed as well as the uniformity and porosity of the barrier and its ability to adhere to the pipe wall. • The solubility of corrosion by-products involving chloride and sulfate needs to be considered for uniform corrosion. • Chemicals, such as orthophosphate, can be added to produce adequate barriers against uniform corrosion on the pipe wall.
Microbiologically Influenced Corrosion
The extent that microorganisms are involved in the chemical and physical phenomena that influence the water quality in a distribution system has not been adequately studied. Nevertheless, there is enough knowledge about microbial growth that microorganisms can be prevented from growing excessively in water systems. Preventing excessive growth of microorganisms in water systems lowers the possibility of microbiologically influenced corrosion of metals.
Accumulation and Transport
• Metals and other inorganic material can accumulate on pipe walls by both gravitational and biofilm-mediated processes. • Lead, copper, and other metals such as arsenic and radium can be sorbed and coprecipitated with the inorganic material and become ingrained in the pipe wall scale. • Accumulated materials, including lead and copper, can be released by both chemical and physical means back into the water at higher concentrations than seen at entry points to the distribution system.
Other Metal Release
There are other mechanisms of metal release that should not be overlooked: galvanic corrosion, concentration cells, erosion corrosion, pipe manufacturing quality, pipe installation, and system maintenance.
Summary
Understanding and addressing all of the mechanisms of metal release is critical to lowering lead, copper, iron, and other metals concentrations in drinking water.

MECHANISMS OF METAL RELEASE BY UNIFORM CORROSION

Importance of a Protective Barrier

The correct way to apply the term *corrosion* to drinking water systems is in reference to an electrochemical interaction between a metal surface, such as a pipe wall, and the water with which it is in contact. The electrochemical interaction is similar to that of a battery (Figure 2-1) (Peabody 2001).

The components of a battery are (Peabody 2001)

- negative terminal called the *anode,*
- positive terminal called the *cathode,*
- medium to carry electrons from the anode to the cathode,
- chemical that can accept the electrons at the cathode, and
- chemical that can pair with the metal ion released at the anode.

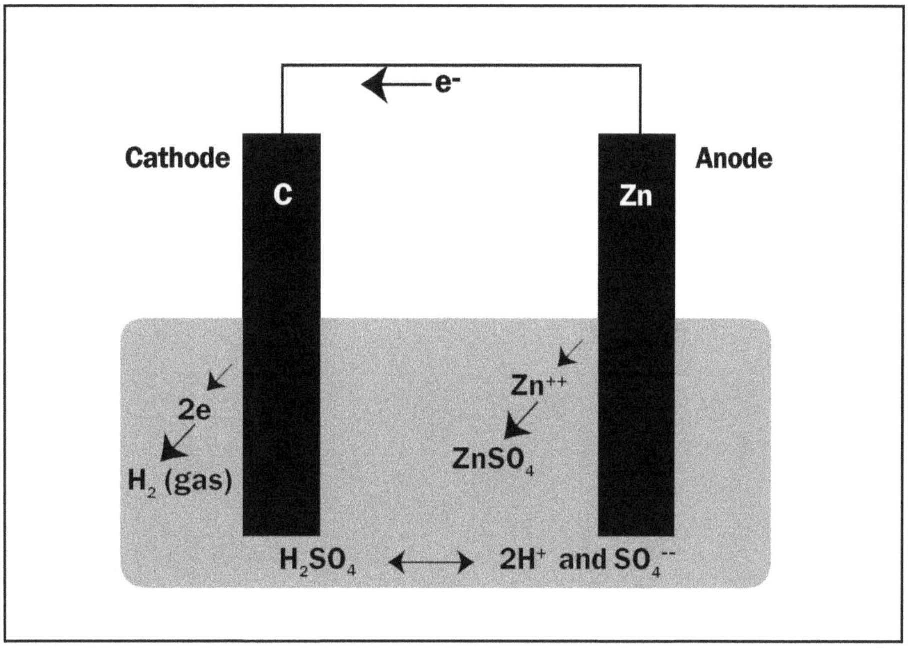

Figure 2-1 Typical battery

In uniform corrosion taking place inside the piping of water distribution systems, the anode and cathode occur dynamically at random sites on the pipe wall (Figure 2-2). These "negative and positive terminals" are microscopic in surface area. They are thought to occur because of natural microscopic imperfections and variations on the metal surface (AwwaRF and TZW 1996).

The electrons flow from the microscopic anodes to the microscopic cathodes through the pipe wall, which is similar to the role that wiring plays in an electrical circuit (AwwaRF and TZW 1996).

The water solution in contact with the cathode provides the chemicals to accept the electrons (AwwaRF and TZW 1996). For instance, oxygen in the water solution can accept an electron, and the reduction reaction occurs:

$$e^- + \frac{1}{4} O_2 + \frac{1}{2} H_2O \rightarrow OH^-$$

In the meantime, metal atoms from the solid metal at the anode, having given up electrons, undergo an oxidation reaction. The oxidized metal is now soluble in the adjacent water solution. That is, solid metal has been lost, transferred to water as a metal ion. It is this oxidation of the metal that is termed *corrosion*. Because of the dynamically changing anodes, there is equal probability that metal will be lost at any site along the pipe wall. Therefore, the term *uniform corrosion* is used (AwwaRF and TZW 1996).

The water solution now contains positively charged metal ions from the anode and negatively charged ions from the cathode as well as various ions from the water source and added water treatment chemicals. New compounds are formed as oppositely charged ion pairs (AwwaRF and TZW 1996).

Figure 2-2 suggests, in a simplified manner, methods to stop uniform corrosion from occurring. There is no wire with an on/off switch to break the flow of electrons. Only two components in the water/pipe system can be manipulated. First, electron acceptors in the water solution, such as oxygen, can be eliminated. This approach is sometimes, but not always, practical in water treatment. A second approach is to create a barrier, or protective (passivating) film, between the metal surface and the water solution (Figure 2-3). However,

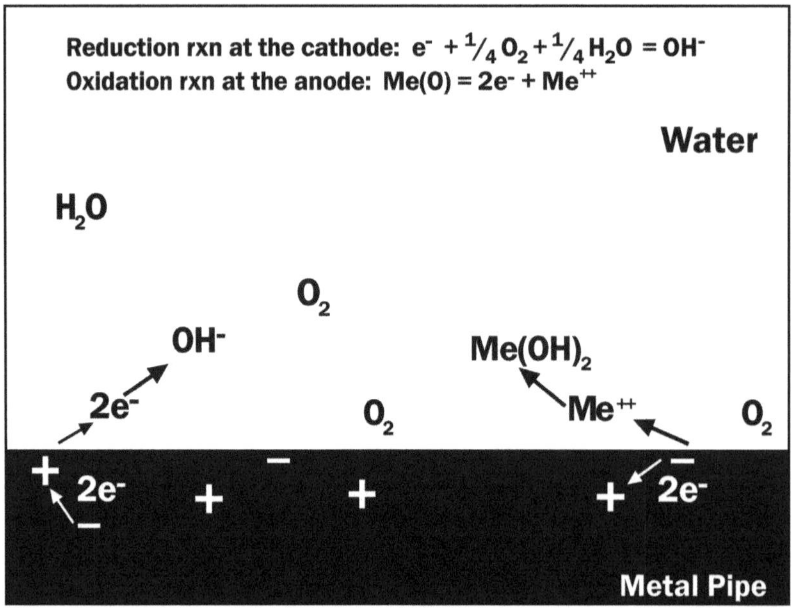

Figure 2-2 Typical electrochemical interaction between water and metal pipe

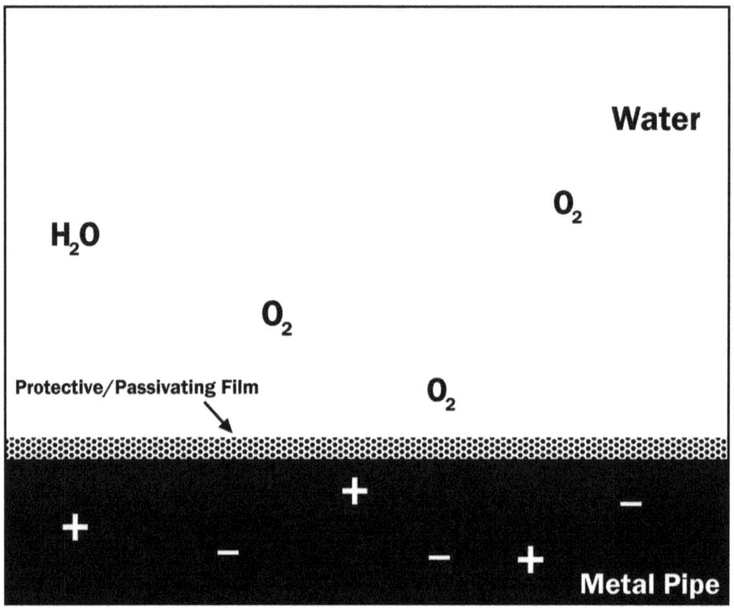

Figure 2-3 Barrier between water and metal pipe

the film must be an insoluble, uniform, nonporous layer that is well adhered to the pipe wall, preventing electron acceptors from contacting the electrons at the cathodic sites on the pipe wall (AwwaRF and TZW 1996). Further details related to forming a passivating film are described later in this chapter in context with various metals commonly used in water piping.

The Failure of Calcium Carbonate Scales as Protection Against Uniform Corrosion

If a barrier can be created between the metal surface and the water, then uniform corrosion can be stopped. However, the barrier must be uniform, nonporous, insoluble, and well adhered to the pipe wall to prevent contact between the metal and the water and prevent the flow of electrons and molecules. As stated previously, a barrier with these characteristics is called a *passivating film*. Many decades ago, it was thought that calcium carbonate could create such a barrier.

Calcium carbonate ($CaCO_3$) naturally occurs in water and is easy to add to water and manipulate in water. It is known that changes to pH, temperature, calcium hardness, and alkalinity can force calcium carbonate to precipitate out of solution. A number of simple methods have been developed to calculate relative quantities of calcium carbonate that will precipitate as a scale. The most common of these methods are the Langelier index, the calcium carbonate precipitation potential, the Aggressiveness index, and the Ryznar index. In the past, the goal has been to achieve a certain quantity of calcium carbonate scale that will stop uniform corrosion but not so much scale that pipes will become significantly reduced in diameter or plugged.

It would be convenient if corrosion control and control of metal release were as simple as calculating the Langelier index. However, an important corrosion reference book, *Internal Corrosion of Water Distribution Systems* (AwwaRF and TZW 1996), emphasizes the inadequacies of calcium carbonate scale. These arguments are based on microscopic studies of this scale's physical properties on lead, copper, and iron pipe walls (AwwaRF and TZW 1996). It has been confirmed that this coarse scale cannot stop the electron receptor molecules from migrating to the pipe wall. Studies cited in the book also show the lack of correlation of Langelier index with concentration of metals in the water (AwwaRF and TZW 1996).

Because calcium carbonate is inadequate as a passivating film, other pipe wall barriers must be considered.

Other Pipe Films and Scales

An attempt to create the perfect physical barrier to uniform corrosion of a metal surface can be made by considering the metal ions that are released into the water from the anodes. It is the physical and chemical characteristics of the new compounds formed with the released metal ions that determine whether or not uniform corrosion will continue, slow down, or stop.

An important characteristic of the new compounds is their solubility in water. If a new compound formed with the metal ions is highly insoluble, the compound will stay in a solid form that will possibly accumulate on the metal surface. If the solid forms a uniform, nonporous film that adheres to the pipe wall, a physical barrier to uniform corrosion can be created. If the new compound formed by metal ions is highly soluble, very little to none of the compound will be found, because solids on the metal surface and the dissolved metal will be carried away by the water. Uniform corrosion will persist at a high rate as long as the components of the soluble compound are provided in the water to interact with the released metal ions (AwwaRF and TZW 1996).

The new compounds formed using the metal ions from the metal surface can be called, "corrosion by-products" or "by-products of uniform corrosion." The reality is that many corrosion by-products form with naturally occurring ions in the water, such as carbonates (CO_3^{2-}), hydroxides (OH^-), and oxygen (O_2). The new compounds are only partially insoluble, so a natural balance occurs between the metal compound in the solid form that accumulates on the metal surface, the metal compound in the solid form that remains entrained in the water, and the dissolved metal compound. Multiple compounds can remain in these various states at the same time. Attempts can be made to push one or more of the compounds into a more insoluble, barrier-forming state by adjusting pH and/or alkalinity or adding a corrosion inhibitor (such as orthophosphate or silicate).

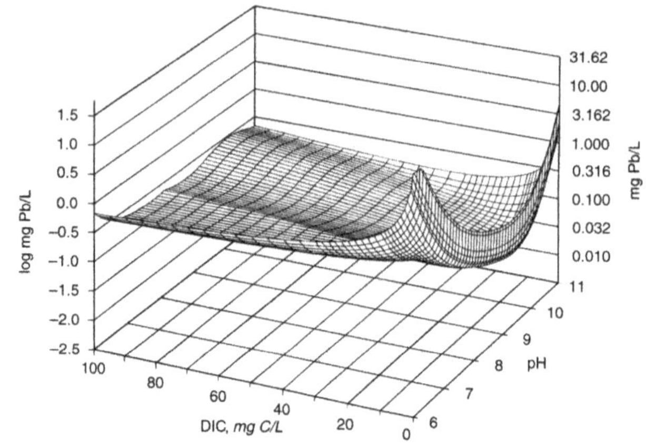

Three-dimensional plot of the relationship of lead solubility as log (mg Pb/L) to pH and DIC concentration at 25°C and I = 0.01 mol/L

Source: *AwwaRF and TZW 1996*

Figure 2-4 Graphical representation of lead carbonate solubility model

The manipulation of corrosion by-products can be illustrated by considering the control of lead, copper, and iron uniform corrosion as discussed below.

Lead

Carbonate. Researchers began their exploration for adequate protective barriers on lead pipe by examining the structure and composition of scales that had built up naturally on pipe walls. A classic example of this type of exploration was described in a 1987 article by Colling and colleagues. The researchers examined the composition of lead pipe wall deposits using X-ray diffraction and infrared spectroscopy. The deposits' structure was viewed with scanning electron microscopy. Other similar studies were performed in Europe and in the United States in the 1970s and 1980s (AwwaRF and TZW 1996).

The most common compounds found on a lead pipe wall are cerussite ($PbCO_3$) and hydrocerussite ($Pb_3(CO_3)_2)(OH)_2$). Researchers' early observations found that "the concentrations of lead [in the water] are controlled by the solubility of lead salts in the pipe deposits" (Colling et al. 1987).

Given that lead pipe deposits were found to have carbonate (CO_3^{-2}) compounds as a major component, dissolved inorganic carbon (DIC) became the focus of subsequent studies. Controlled experiments with lead pipes were devised. DIC and pH were varied in the water, and the resultant lead concentration was studied (Colling et al. 1987, AwwaRF and TZW 1996).

Taking this relationship a step further, a theoretical lead carbonate solubility model was developed. Thermodynamic solubility constants were used to calculate the lead concentration in water based on DIC and pH (Schock 1980). Figure 2-4 is a graphical representation of the lead carbonate solubility model, where soluble lead concentration in water is predicted by DIC and pH. This lead carbonate solubility model became the basis for directives to water utilities for lead control in the Lead and Copper Rule (LCR), part of the federal primary drinking water regulations in the United States when the rule was published in 1991 (*Federal Register* 1991).

Oxide. As more experience was gained by water quality personnel working to comply with the LCR, it became obvious that the lead carbonate model did not predict all releases of lead into drinking water. First, the model did not consider compounds formed when lead was in the +4 oxidation state. In this state, lead has lost four electrons in the corrosion process instead of two, as was only considered for the model shown in Figure 2-4. In the higher oxidation state, lead can form an oxide called plattnerite (β-PbO_2) or its polymorph, scrutinyite (α-PbO_2) (Lytle and Schock 2005). These compounds are highly insoluble in water and can form the desired passivating barrier of solids on metal surfaces. The phenomenon has been seen to greatly slow the uniform corrosion of lead with resultant low-lead concentrations in drinking water (Lytle and Schock 2005, Wang et al. 2012).

The effects of lead in the +4 oxidation state came into focus around 2004 when observations were made during a change in disinfection chemicals from chlorine to chloramine at one water utility. Water utilities can choose to use chloramine to maintain disinfection throughout a water distribution system rather than chlorine to lower the possibility of creating carcinogenic disinfection by-products. Because chloramine is not as strong an oxidant as chlorine, the oxidation–reduction potential (ORP) of the water is lowered after this chemical change. This means that lead released in the corrosion process can no longer achieve the highly oxidized state needed to form the relatively insoluble plattnerite/scrutinyite film. The existing protective film is lost, and the lead concentration in the water can increase greatly (Lytle and Schock 2005). Because of this finding, it is now recognized that keeping a high ORP in water systems to form plattnerite/scrutinyite on lead surfaces can help control lead.

Hydroxide. Under certain conditions that are not common for drinking water characteristics, lead hydroxides can play a significant role as a corrosion by-product. One utility's water historically has had a very low alkalinity with a very high pH of over 10. The water utility was encouraged to lower the pH of the water to below 10 based on the predictions of the lead solubility model represented in Figure 2-4. However, the lead concentration increased. Further investigation determined that the water system had a significant presence of a compound on its pipe walls not predicted by the model. It was a type of lead hydroxide called plumbonacrite, where its solubility decreases as the pH is increased and approaches 13. When the pH was lowered to match the lead solubility model's predictions of Figure 2-4, the plumbonacrite re-solubilized and increased the lead concentration in the water (Rego and Schock 2007). This emphasizes the need to understand the type of lead compounds actually present in the system and not just those predicted by the lead solubility model of Figure 2-4.

Chloride and sulfate. Chloride and sulfate can be present naturally in drinking water or from contamination of water sources, such as with the influence of road salt on well water. The source of chloride and sulfate in drinking water can also be from treatment chemicals and operational changes. For example, chloride and sulfate concentrations are affected by switching from chlorine gas to hypochlorite generation using brine solutions—changing from a sulfate-containing coagulant such as alum to chloride-containing coagulant such as ferric chloride—anion exchange for arsenic treatment, and desalination.

In terms of lead corrosion, the solubilities of lead chloride and lead sulfate are many orders of magnitude greater than that of lead carbonate. Table 2-2 shows a comparison. That is, if high levels of chloride or sulfate are found in water, high dissolved lead chloride or sulfate compound concentrations would be expected.

It is important to keep in mind that the effects of chloride and sulfate will vary based on the ratio of chloride-to-sulfate concentrations in the water *and* the ratio of alkalinity to the sum of the chloride and sulfate concentrations (Nguyen et al. 2011). The carbonate ions, as expressed by the measurement of alkalinity, compete with the chloride and sulfate. A high concentration of carbonate ions (a high alkalinity) can passivate lead when the chloride and sulfate concentrations are relatively lower (Larson and Skold 1958, Nguyen et al. 2011). However, a higher chloride-to-sulfate mass ratio can increase galvanic corrosion (corrosion at the interface of adjacent dissimilar metals) and subsequently increase lead

Table 2-2 Solubility of various lead compounds found in drinking water

Compound	Chemical Formula	Solubility in Grams of Compound per 100 g Water at 20°C
Lead carbonate	$PbCO_3$	0.00011
Lead sulfate	$PbSO_4$	0.0041
Lead chloride	$PbCl_2$	0.99

Source: *Perry and Green 1984*

release from leaded components (Edwards and Triantafyllidou 2007, Nguyen et al. 2011). Therefore, the relationship among alkalinity, chloride concentration, and sulfate concentration should be investigated in each water system under study.

Orthophosphate. So far, this discussion has described scales and films formed by compounds of naturally occurring ions such as CO_3^{-2}, O^{-2}, and OH^-. There are also chemicals that can be added to water to intentionally form passivating films on metal surfaces to minimize uniform corrosion of lead.

One common chemical is orthophosphate (PO_4^{-3}). Its ability to form passivating films on lead surfaces in contact with water uses the very same mechanisms previously discussed. That is, metal ions are released into the water from the anodic sites on the metal surface. The metal ions form compounds (corrosion by-products) with the phosphate ions that have been added to the water. The lead–phosphate compound theoretically precipitates out of the water as a solid and forms a film on metal surfaces to prevent contact of the pipe with the water to minimize uniform corrosion of lead (AwwaRF and TZW 1996).

Several lead and phosphate compounds are possible, but research has concluded that a compound called hydroxypyromorphite ($Pb_9(PO_4)_6$ or $Pb_3(PO_4)_2$) is most likely the solid that controls the solubility of lead in the water when orthophosphate has been added (AwwaRF and TZW 1996). Many technical articles demonstrate the effectiveness of orthophosphate in offline pipe test rigs for lowering lead released to water. It has also been concluded that the solubility of the lead and orthophosphate compounds that form is dependent on pH and DIC in addition to the concentration of orthophosphate (AwwaRF and TZW 1996).

The USEPA recommends the use of orthophosphate as phosphoric acid or a potassium or sodium salt and not other products such as zinc orthophosphate or blends with polyphosphate as described below (USEPA 2016).

In another consideration, despite the LCR's dependence on the use of orthophosphate for lead and copper corrosion control in drinking water systems, the regulation of phosphorus discharged by wastewater treatment plants around the country is increasing for nutrient control in receiving waters. In 2016, twelve states have wastewater phosphorus limits in place as regulated by US National Pollutant Discharge Elimination System permits (Rodgers 2014). These Clean Water Act regulations along with the requirements of the Safe Drinking Water Act's LCR must be considered for each drinking water system in the United States and similar environmental balances considered elsewhere.

Zinc orthophosphate. The use of zinc orthophosphate to minimize lead in drinking water has caused much confusion. It is theorized that zinc forms carbonates that block cathodic sites on the pipe wall while phosphate forms compounds blocking anodic sites (AwwaRF and TZW 1996). There may be advantages for use of this chemical for galvanized iron, where the zinc lining is replenished thereby inhibiting release of accumulated lead (Grasha et al. 1981), and for cement mortar–lined piping, where calcium leaching is inhibited (AwwaRF and TZW 1996). However, there is no evidence that zinc orthophosphate offers more benefit for lead control with other piping materials than other orthophosphate compounds (Schneider et al. 2011).

Also, zinc can cause environmental problems for wastewater treatment plants in terms of finished water discharge and sludge disposal. These environmental factors should be considered when weighing chemical options for corrosion control.

The use of this chemical is discouraged by USEPA guidance (USEPA 2016).

Polyphosphate and blends. Polyphosphate–orthophosphate blends have been used with the intent to create corrosion control barriers on pipe walls. Polyphosphates are polymers composed of repeated linked units of orthophosphate ions in various molecular structures. They have completely different properties than simple non-polymeric orthophosphate compounds such as phosphoric acid (H_3PO_4) or potassium orthophosphate (K_3PO_4). How the phosphate ions in orthophosphate compounds can form new compounds with lead ions that have low solubility in water was previously discussed. The new compounds theoretically precipitate out of the water and form a protective barrier on the pipe wall.

In contrast, polyphosphate compounds are known to bond with metal ions, bringing them into their molecular structure and holding these metal ions in solution. That bonding feature is the very reason that polyphosphates have traditionally been used in water distribution systems to "sequester" calcium, iron, and manganese. That is, calcium, iron, and manganese ions in the water are taken up by the added polyphosphates and not allowed to form other compounds where they might precipitate as solids on water customers' sinks and laundry. Although it was known many years ago that calcium, iron, and manganese kept in solution were sequestered, it was not realized that other metal ions in the water, such as lead and copper, were also captured. Even if the polyphosphate is blended in a product with orthophosphate, this polyphosphate–metal mixture can be consumed by the water utility customer and can be measurable in regulatory samples for lead and copper (Holm and Schock 1991, Cantor et al. 2000).

Since the publishing of the LCR in 1991, polyphosphates and blends of polyphosphates and orthophosphates have been marketed as a means to both sequester iron and manganese and minimize lead and copper in the system by precipitation of a fine metallic compound as a barrier on the pipe wall. Polyphosphate products and blended polyphosphate–orthophosphate (poly-orthophosphate) products can perform these two opposing functions but not necessarily at the same time (Klueh and Robinson 1988, Koudleka et al. 1982, Zinder et al. 1984). When the polyphosphate product is first introduced into the distribution system, it is in its polymeric form and sequesters metals, including iron, manganese, copper, and lead. However, a number of factors in the distribution system environment break up the polyphosphate compound into orthophosphate ions. These factors include pH, temperature, time, polyphosphate concentration, polyphosphate-to-orthophosphate ratio, type of polyphosphate species, and the presence of other various constituents in the water. When the polyphosphate has reverted to orthophosphate ions in this way, it can have the passivating effect on the pipe wall described previously (Klueh and Robinson 1988, Koudleka et al. 1982, Zinder et al. 1984).

A problem with adding polyphosphate as an agent to minimize lead and copper release is that there is no way to predict where and when the polyphosphate will revert to orthophosphate. Even if a prediction could be made, there would most likely be some locations and time frames in which water utility customers were consuming the higher metal concentrations until the reversion to orthophosphate could take place.

Past technical literature has warned against using polyphosphates (AwwaRF and TZW 1996). One study showed lead concentration increasing to four times that of untreated water with the addition of a polyphosphate–orthophosphate blend (Cantor et al. 2000). Holm and Schock (1991) concluded that "the application of polyphosphates for the specific purpose of lead corrosion control entails considerable uncertainty and risk."

USEPA guidance discourages the use of phosphate products that contain polyphosphate compounds and recommends stringent demonstrations of such a product's effects on lead and copper release from piping material (USEPA 2016).

Sodium silicate. Sodium silicate is sometimes used as an agent to minimize lead concentrations in water; however, little research and field information are available to prove its effectiveness. The limited work that has been done has shown that a relatively high concentration of silicate is needed (around 20 mg silicon dioxide (SiO_2)/L) and that it may take months to see any reduction in lead concentrations (AwwaRF and TZW 1996). In addition, sodium silicate is successfully used to sequester iron and manganese—just like the dual use of polyphosphates. It is not known by what mechanism sodium silicate works both to prevent lead from entering the water and to sequester iron and manganese. It may form a passivating compound on the pipe wall or it may adsorb lead, iron, and manganese ions to the surface of a silicate film on the pipe wall (AwwaRF and TZW 1996). It may also act to increase the pH of low-alkalinity water that, in turn, would lower the uniform corrosion potential for lead and copper (AwwaRF and TZW 1996). Sodium silicate has been mostly eliminated from many utilities' plans of action because of this lack of information, the increased operating cost associated with the cost of the chemical and increased dose, and the time it takes to see results. A 2005 article brought the possibilities of sodium silicate use back to life by describing both successful reduction of lead and copper and increased aesthetic qualities in terms of iron and manganese in a water system (Schock et al. 2005).

Free chlorine. Free chlorine, used as a disinfectant in water distribution systems, is a strong oxidant. One would expect that a strong oxidant would readily accept electrons at the cathodic sites on the pipe, and corrosion would increase at the anodic sites (AwwaRF and TZW 1996). However, that assumption is not always correct. As was discussed previously, when the presence of free chlorine pushes the ORP of the water high enough, lead is able to lose four electrons instead of two (AwwaRF and TZW 1996). When lead loses four electrons, it can form a highly insoluble and protective lead dioxide (PbO_2) in either of two forms: plattnerite or scrutinyite. Lead concentration in the water is, by consequence, lowered. Refer to the "Oxide" section previously.

Conclusion. This discussion on the uniform corrosion of lead demonstrates that the release of a metal, such as lead, into water is a complex combination of competing chemical interactions, microbiological interactions (see the "Mechanisms of Metal Release by Microbiological Activity" section later in this chapter), and transport by accumulated pipe wall scales (see the "Mechanisms of Metal Release by Accumulation and Transport" section later in this chapter). Every situation of elevated lead has its own nuances that must be understood. The lead solubility model of Figure 2-4 does not and cannot express these multiple factors and predict the lead concentration in water. Instead, an attempt at uncovering the significant multiple factors of lead release must be made. This Manual of Water Supply Practices provides the knowledge base necessary to perform such an exploration.

Copper. In considering the uniform corrosion of copper, the identification of mechanisms that encourage the metal's release to water is just as complex as that for lead. There are many in the field of internal corrosion control that see uniform corrosion based on copper carbonate and oxide compound solubility as the primary cause of observed copper concentrations in excess of the LCR's action level of 1,300 μg/L. In addition, others have observed many situations where uniform corrosion based on copper chloride and sulfate compound solubility predominates and where microbiologically influenced corrosion predominates (Cantor 2014).

Carbonate, oxide, hydroxide, and free chlorine. Copper release is high when new copper pipe first comes in contact with water, and the young scales are predominately cupric hydroxide ($CuOH_2$) (Schock et al. 1995, Edwards et al. 1996). Over time, the least soluble copper compounds, such as malachite ($Cu_2CO_3(OH)_2$, a copper carbonate) and tenorite (CuO, a copper oxide) build up on the metal surface and a more protective barrier develops. Copper release to the water slows and copper concentration in the water drops. The rate and degree to which this will occur is dependent on pH, alkalinity, and chemicals in the water that may interfere with scale development. Some situations where only chemical factors were measured were observed to reach passivation in four days and others over 10 years (Hidmi and Edwards 1999, Turek et al. 2011). A 2014 presentation (Schock and Lytle 2014) explained the chemistry behind uniform corrosion of copper based on carbonate compound solubility. In this case, alkalinity and pH were described and represented graphically as the major factors affecting the uniform corrosion of copper.

The copper solubility graph (Schock and Lytle 2014) from the 2014 presentation describes that the following conditions are prone to copper release greater than the LCR action level of 1,300 μg/L by uniform corrosion:

- pH <7
- pH 7 to 7.2 and alkalinity >35 mg/L as $CaCO_3$
- pH 7.2 to 7.5 and alkalinity >100 mg/L as $CaCO_3$
- pH 7.5 to 8 and alkalinity >150 mg/L as $CaCO_3$
- pH 8 to 8.3 and alkalinity >250 mg/L as $CaCO_3$

In addition, ORP is a major factor in shifting young copper hydroxide scales into a soluble phase (Schock and Lytle 2014). ORP is influenced by the concentration of chlorine disinfectant in the water because it is an oxidant. With this information, the researchers concluded that any chlorinated water system is prone to excessive copper release by uniform corrosion.

Alternatively, others have found that the pH, alkalinity, and ORP relationships to high copper release may not always predominate and that other factors such as microbiologically influenced corrosion (MIC), chloride, and sulfate may be more important in many situations (Cantor 2014).

Chloride and sulfate. The technical literature regarding chloride and sulfate and copper corrosion is contradictory. In some technical literature, the presence of chloride formed a passivating scale (Edwards et al. 1994) and lowered the copper corrosion current (Edwards et al. 1996) while in other literature, the presence of chloride increased the soluble copper concentrations over the predicted amount (Schock et al. 1995). Others found that chloride initiated copper pitting (Ives and Rawson 1962, Edwards et al. 1994).

Natural organic matter. Natural organic matter (NOM) has been reported to increase copper solubility by forming soluble complexes with copper and interfering with protective scale formation (Schock et al. 1995, Korshin et al. 1996, Edwards and Sprague 2001). Others found that NOM prevented copper pitting (Edwards et al. 1994, Edwards and Sprague 2001, Sarver and Edwards 2012). As discussed later in this chapter in the "Mechanisms of Metal Release by Microbiological Activity" section and in chapter 3, the presence of NOM can also act as a food source for microorganisms with higher levels of NOM, increasing microbial activity and the occurrence of MIC.

Orthophosphate. Orthophosphate is effective in lowering copper release into water from uniform corrosion where the pH is above 7.0. At acidic pH values, orthophosphate is ineffective in lowering copper release (AwwaRF and TZW 1996). At higher pH values around 8.0 where copper release is diminished because of the pH increase, orthophosphate addition shows no additional benefit (AwwaRF and TZW 1996, Schock and Sandvig 2006). Orthophosphate prevents high copper release in alkalinities greater than 200 mg/L as $CaCO_3$ on new pipe in test apparatuses, if dosed with at least 3.3 mg/L as orthophosphate (Shock and Lytle 2014).

Polyphosphate and blends. The previous discussion concerning lead pointed out how the use of polyphosphate and some polyphosphate–orthophosphate blends to mitigate uniform corrosion often has the opposite effect. It is the polyphosphates' reversion to orthophosphate that gives it the ability to minimize metal release if conditions are favorable. Many studies reported in the literature have shown that copper levels can increase dramatically with the addition of polyphosphate and blends (AwwaRF and TZW 1996). Studies carried out on polyphosphate-blend addition in hot water showed a benefit in reduction of copper release, but it was also acknowledged that the higher temperatures push the polyphosphate to revert to orthophosphate (AwwaRF and TZW 1996).

One case study by Knobeloch et al. (1998) exemplifies the point that the use of polyphosphate–orthophosphate blends can be unpredictable. Some water customers in one city experienced staining of sinks and hair and fell ill with gastrointestinal upsets. Their malaise was traced to high copper levels (>5,000 μg/L) in the drinking water (Knobeloch et al. 1998). A polyphosphate–orthophosphate blend had recently been added to sequester dissolved iron in the groundwater source. An investigation proved that by removing the polyphosphate blend from the drinking water, copper levels in new homes were dramatically lowered (Cantor et al. 2000). This case study is related here to demonstrate what can happen when polyphosphate, either alone or as a polyphosphate–orthophosphate blend, is added to a drinking water system and is not properly monitored. The literature is clear that metals will increase in concentration in the presence of polyphosphate until the polyphosphate reverts to orthophosphate.

Sodium silicate. As previously discussed with lead, there is not enough literature to assess the effectiveness of sodium silicate as a means to provide a protective barrier between metal and water. Studies have not determined whether the silicate decreases the release of copper into the water or if the increase in pH as a result of sodium silicate addition gives the benefit of copper control (AwwaRF and TZW 1996).

Conclusion. As in the case of lead release from metal surfaces, copper release is a complex interaction of multiple chemical and microbiological factors and transport by pipe wall scales. The copper solubility model, which describes a simplified view of uniform corrosion based solely on carbonate chemistry, cannot predict, by itself, the copper concentration in water.

Iron. Iron concentration in drinking water has not historically been a concern from the perspective of regulatory compliance because it has typically been identified as an aesthetic concern in water. However, as will be discussed later in the "Mechanisms of Metal Release by Accumulation and Transport" section of this chapter, iron can sorb lead and other metals, accumulate them in scales, and transport them to the consumers' taps. Therefore, the production of iron particulates as a result of the corrosion of iron plumbing materials can eventually lead to a major mechanism of lead release into the drinking water (Schock et al. 2014). Therefore, iron corrosion in a water system must be controlled with the same chemical and microbiological concerns as with lead and copper.

Alkalinity/DIC and pH. As with lead and copper exposed to water, iron exposed to water is affected by oxygen and other oxidants as electron acceptors in the electrochemical battery of metal corrosion. And once again, alkalinity and pH play a major role in forming compounds with the metal ions that are released into the water (AwwaRF and TZW 1996).

The corrosion scale for iron pipe is quite complex and can follow several pathways as to which compounds are formed, whether or not they increase or decrease the corrosion of the iron piping, and whether or not they remain soluble in the water or form solids (rusty water). If there is a slow oxidation of the metallic iron, then the resultant corrosion by-product formation moves toward the production of $FeCO_3$ (siderite)—for example, a relatively nonporous scale that can hinder the migration of oxidants to the cathodic sites on the pipe wall. But if there is rapid oxidation of the metallic iron, a less dense and more porous iron scale is formed. In this case, oxygen and ions can migrate to the pipe wall through the scale as well as react with compounds in the scale itself. Other factors, such as chloride and sulfate concentrations, can play a role in further transformation to other iron compounds (AwwaRF and TZW 1996).

Summaries of the physical and chemical characteristics of the layers and factors that influence iron release are described elsewhere, as in an article by Burlingame et al. 2006 and in chapter 2 of AwwaRF and TZW 1996.

Orthophosphate. Orthophosphates have been shown to reduce the corrosion rates of iron and also to stabilize the iron scale that forms on the pipe surface (AwwaRF and TZW 1996).

Polyphosphate and blends. Polyphosphates and blends have been shown to increase iron uniform corrosion rates and destabilize the iron scale on the pipe surface (AwwaRF and TZW 1996). That is to say, the same chemical mechanism that allows polyphosphate to be used as a sequestering agent for iron actively pulls iron scale from pipe surfaces and holds the iron in solution, thereby increasing iron concentration in the water.

Free chlorine. Observations where only chemical parameters were measured have been made that free chlorine increases the release of iron into the water (Cantor et al. 2003). In addition, the physical characteristics of the iron scale depend on the rate of reaction between the iron and the oxidant. Free chlorine reacts rapidly, with soluble iron forming more-porous and less-dense solids than those formed in a slower reaction with oxygen or chloramine (Burlingame et al. 2006).

Conclusion. In terms of lead and copper control, iron corrosion is important to track in a water system because iron corrosion by-products can facilitate lead and copper transport to consumers' taps. This is discussed later in the "Mechanisms of Metal Release by Accumulation and Transport" section. Iron is also susceptible to MIC, discussed in the next section. Like lead and copper, iron undergoes complex chemical and microbiological interactions in water systems.

MECHANISMS OF METAL RELEASE BY MICROBIOLOGICAL ACTIVITY

Microbiologically influenced corrosion (MIC) is a type of nonuniform corrosion. However, it should not be necessarily considered as a localized phenomenon because the conditions that cause this type of corrosion to occur can be pervasive throughout a water system.

MIC has been greatly overlooked as a cause of lead and copper (and iron) release into drinking water in the literature, which is likely related to the difficulty in quantifying the presence of microorganisms in the past. With improvement in analytical techniques, investigators who measure microbiological populations along with chemical interactions in water systems observe that MIC can not only elevate metals concentrations in the adjacent water but can also cause pitting and pinhole leaks starting from the interior surface of a metal pipe or plumbing component (Little et al. 1992, Bremer et al. 2001, Fallowfield et al. 2011, Cantor 2014).

Also with the advance of technology, a new perspective of water systems is emerging with the realization that water distribution systems have a microbiome, a community of microorganisms that thrives in the system (Raskin et al. 2014, Gomez et al. 2014, Ji et al. 2014, Douterelo et al. 2014). At this time, little is known about the extent that microorganisms become intertwined in the chemical and physical phenomena that shape the water quality in a distribution system.

Nevertheless, some straightforward measurements can aid in tracking and controlling excessive growth of microorganisms in water systems. These simple concepts alone can lower the possibility that MIC and its associated water quality and pipe integrity problems will occur.

Microbial Growth

The term *microbial growth* is a general descriptor of the process used by microorganisms to replicate, which generally involves bulking up in size and then dividing to make additional microbial copies. For microbial growth to occur in water, certain macronutrients, chemical components, and micronutrients must be present for use by microbes. These chemicals provide the raw materials and energy for tearing down and reassembling cellular materials.

Macronutrients. Macronutrients are the chemical compounds consumed in the largest quantities by microorganisms that contribute the most energy to power their life cycles. The generic chemical formula for microorganisms is roughly $C_5H_7O_2NP_{0.05}$. Therefore, required macronutrients for microbial growth include carbon (C), phosphoros (P), nitrogen (N), hydrogen (H), and oxygen (O) (Rittmann and McCarty 2001).

Organic carbon is the preferred form of C for microbial growth because it allows for the fastest growth rate. The type of organic carbon present provides the potential for particular types of microbial growth. Typical forms of organic carbon utilized by microbes include

- organic acids (e.g., acetic, citric, and formic);
- alcohols (e.g., ethanol, glycerol, and methanol);
- sugars, also referred to as carbohydrates (e.g., glucose, pyruvate, and succinate);
- amino acids (e.g., alanine and glutamate); and
- complex organic decay products (e.g., lignin, tannin, and cellulose) (Rittmann and McCarty 2001).

Organic carbon in surface water is typically the result of decaying vegetation and/or aquatic organisms in the water, which is often defined as natural organic matter, or NOM. This can be augmented by organic contaminants (e.g., animal feces, fertilizers, and petroleum products) that are transported by stormwater runoff or from sewage system inputs (e.g., leaky pipes, outfall, overflows, or failed septic systems) (Ehrlich and Newman 2009, Edzwald 2011). Decay of NOM or organic contaminants can release C, N, and P to the environment and thus stimulate microbial growth (Ehrlich and Newman 2009). However, as discussed later in chapter 3, not all NOM is available for microbial use.

Groundwater recharge traveling through the soil zone will pick up and transport organic carbon toward the aquifer, with organic carbon concentration typically decreasing with depth in the aquifer because of continued microbial degradation (Ehrlich and Newman 2009). Most organic carbon in groundwater is the result of decaying vegetation. In deep aquifers, organic carbon is typically either in a complex form and unavailable for microbial use or the remnant of fossil matter (Goldscheider et al. 2006). Surface water that, for various reasons, can short-circuit the flow into deeper aquifers has the potential for increasing the organic carbon content of groundwater.

However, lack of a suitable organic carbon source does not stop microbiological growth. Certain microbes (autotrophs) can utilize DIC as their source of C instead of organic carbon (Goldscheider et al. 2006, Rittmann and McCarty 2001).

Requirements for P are typically satisfied by dissolved inorganic phosphate. Various microbes appear to have the ability to acquire inorganic phosphate that is either bound to an oxidized metal or within a crystalline matrix (mineral weathering). Organically bound phosphorus can also be a source of P (Jaisi et al. 2011, Rogers and Bennett 2004).

Ammonia is the preferred form of N. However, certain microbes have the ability to utilize nitrate, nitrite, and nitrogen gas for their growth needs (Rittmann and McCarty 2001).

Requirements for H and O are typically satisfied by acidity (lower pH water has a higher concentration of hydrogen ions) and carbon dioxide or dissolved oxygen, respectively (Rittmann and McCarty 2001).

Chemical components. Nearly every living cell uses adenosine triphosphate (ATP) as its basic energy carrier (Wolfe-Simon et al. 2011), which is used to support processes needed for the assimilation of nutrients into new cellular materials. The ATP energy cycle requires chemical components for microbial growth that include an electron donor and an electron acceptor for energy generation (Rittmann and McCarty 2001).

The use of organic carbon as the electron donor classifies a microbe as a *heterotroph,* whereas the use of an inorganic compound or element as the electron donor (e.g., dissolved metal, sulfide, ammonia, methane, or hydrogen) classifies a microbe as an *autotroph* (Rittmann and McCarty 2001).

The use of oxygen as the electron acceptor classifies a microbe as an *aerobe,* whereas the use of an inorganic compound or element as the electron acceptor (e.g., nitrate, sulfate, oxidized metal, or carbon dioxide) classifies a microbe as an *anaerobe* (Rittmann and McCarty 2001).

The use of organic carbon as both the electron donor and electron acceptor classifies a microbe as a *fermenter* (Rittmann and McCarty 2001).

Facultative microbes can use multiple forms of electron donors or electron acceptors, whereas *obligate microbes* can only use one form (Rittmann and McCarty 2001).

As an example of the complexity of microbial growth, for growth of autotrophic microorganisms to occur, dissolved carbon dioxide or carbonate and bicarbonate (DICs) satisfies the macronutrient C requirement, and a dissolved metal (such as a reduced form of iron, manganese, sulfide, and/or arsenic), ammonia, nitrogen gas, or methane satisfies the electron donor requirement. The electron acceptor requirement can be satisfied by dissolved oxygen for aerobic conditions or an oxidized metal, nitrate, sulfate, carbon dioxide, or organic matter for anaerobic conditions (Goldscheider et al. 2006).

In summary, there is adaptability and synergy of microbiological communities in obtaining the chemical components and compounds to support growth.

Micronutrients. Micronutrients are chemical compounds required in small quantities to support various cells functions. Examples of micronutrients or trace elements are sodium, potassium, magnesium, calcium, iron, sulfur, manganese, cobalt, zinc, copper, and molybdenum (Rittmann and McCarty 2001, Madigan and Martinko 2006). These chemicals are typically found in sufficient quantities in drinking water.

Microbial growth and energy. As indicated previously, macronutrients (C, N, P, H, and O), chemical components (electron donor and electron acceptor for energy production), and micronutrients are required for microbial growth to occur. The greater the energy difference there is between the electron donor and the electron acceptor, the greater the potential there is for microbial growth and the faster the rate of growth will be (Rittmann and McCarty 2001, Madigan and Martinko 2006).

Under aerobic conditions in the presence of organic carbon, about 30 percent of the biochemically available carbon would be used to provide energy for microbiological growth, with 70 percent of the carbon incorporated into new cellular material. Under these same nutrient conditions, but lacking oxygen (anaerobic), about 50 percent of the organic carbon would be used for energy production. This leaves less carbon available for incorporation into new cellular materials. Therefore, a slower growth rate occurs under anaerobic conditions than aerobic conditions (Rittmann and McCarty 2001).

Lacking the presence of organic carbon, less energetic forms of electron donors are utilized (ammonia, H_2, or sulfide) by autotrophic bacteria, with a large portion of the energy in these compounds used to make organic carbon from carbon dioxide (carbon source) for assimilating into new cellular materials, and thus a slower growth rate (Rittmann and McCarty 2001).

The presence of biofilm, the carbohydrates excreted by some microorganisms to adhere to surfaces and for colony protection, creates the potential for a diverse microbial population in proximity to one another. This could enhance the microbial growth because niche bacteria supplement organic carbon and ammonia, respectively, through processes of carbon fixation and nitrogen fixation (Rittmann and McCarty 2001).

Corrosion Pathways

There are various pathways by which MIC can occur.

MIC via transport of nutrients into cells. For microbial growth to occur, nutrients and required chemical components must cross the microbial cell membrane and enter the cell to satisfy cellular needs. Otherwise, they must be accessible to the outer cell membrane for surface-related reactions (Madigan and Martinko 2006).

Transport of nutrients across the cell membrane is highly regulated by biochemical sensors in the membrane to prevent the unwanted transport of harmful compounds into the cell, and to maintain charge balance, salinity, and pH within the cell.

Under nutrient-rich conditions, nutrients are typically able to freely flow in the direction of the concentration gradient and into the cell through *gated porins* (ion-selective channels) in the cell membrane (Madigan and Martinko 2006).

Under nutrient-poor conditions, nutrients must be actively transported into the cell against the concentration gradient. This requires the microbe to produce proteins bound to the outer membrane of its cell that can capture and transport specific nutrients. The energy used to power the transport is in the form of ATP, which also provides other energy needs of the cell (Madigan and Martinko 2006).

The ABC transport system is the most common form of active transport across the cell membrane. ABC stands for ATP-binding cassette, where a *cassette* is a mobile genetic component. Proteins used for nutrient transport generally have a lock and key arrangement, and they protrude from the exterior surface of a microbe in search of a specific nutrient or compound. Once a protein locks on to a target compound, it is transported across the cell membrane (Madigan and Martinko 2006).

The target nutrient compound could be free floating, loosely sorbed to other compounds, or bound to solid minerals. Studies of mineral weathering in groundwater have shown the ability of microbes to selectively remove iron or phosphorus from solid quartz minerals to satisfy cellular needs (Rogers and Bennett 2004). This suggests that the active transport of nutrients across the cell membrane could cause MIC in water systems if the target nutrient is part of a surface on engineered materials in the water system. That is, the transport mechanism of supplying microbiological cells with nutrients may play a role in MIC of plumbing materials.

MIC via biofilm material. *Biofilm* is the term used to describe the mass (community) of microorganisms and the enzymes and other extracellular material that the microorganisms excrete that allow them to stay attached to a surface. Biofilms are generally described by their makeup of polysaccharides, which can change in response to environmental conditions. Biofilms can be corrosive to pipe surfaces if organic acids are present in the polysaccharides that are excreted by the microorganisms. That is, localized low-pH environments created by the biofilm on the pipe wall can corrode the metal (Bremer et al. 2001).

In addition, chemical reduction and oxidation cycles could occur within the biofilm as symbiotic types of bacteria are brought into proximity. These chemical cycles would affect the state of adjacent metals. For example, the anaerobic heterotrophs in the oxygen-starved interior of the biofilm would convert iron pipe to soluble ferrous iron. Then the aerobic autotrophs on the exterior of the biofilm would convert soluble ferrous iron to solid iron precipitates.

MIC via microbiological products. Biofilms can also be corrosive to pipe surfaces if biological processes occurring in the biofilm result in corrosive environmental conditions (e.g., as with the production of a high-sulfide concentration).

Potential for Occurrence of Microbiologically Influenced Corrosion

Regarding MIC, there are many types of microorganisms to be concerned about in a water distribution system, including but not limited to those that cause nitrification. Potential sources of microbiological nutrients, such as nitrogen, phosphorus, sulfate, and carbon, cannot be ignored (Srinivasan and Harrington 2007, Fleming et al. 2005, AWWA 2006).

Several aspects of a water system environment can be investigated to determine whether MIC is occurring.

Biostability. Biostability of water is a balance of factors that encourage the growth of microorganisms versus factors that discourage their growth. Literature on biostability is available (Van der Kooij 1992, Volk and LeChevallier 2000, Escobar and Randall 2002, Zhang et al. 2002). Factors include macronutrient concentrations (carbon, phosphorus, and nitrogen) in the water, disinfection concentration in the water, and resultant microbiological populations. A simple, but effective method of assessing biostability is to track trends of these factors.

Especially important in that exercise is measuring and tracking the population of all microorganisms in the water at various locations in the water system. A relatively new technology to measure the concentration of ATP in the water produces results that can be translated into the population of microorganisms at the sampling location. As discussed previously, ATP is the energy molecule used by living organisms. The analysis protocol ensures that only ATP from living microorganisms is captured in the sample fraction that is analyzed. If ATP is found in a drinking water sample by this analysis, then there are living microorganisms in the water. The ATP test is a better investigative tool for biostability than older analytical methods, such as heterotrophic plate count. This is because the ATP concentration quantifies all microorganisms in the water (except viruses), not just particular types of bacteria, and is not limited only to bacteria. The specifics of water analyses are explained in chapter 3, as there are several interesting new microbiological analytical techniques. With the ATP test, trends of microbiological populations can be compared with nutrient and disinfection trends. As disinfection is increased or nutrients decreased in water system operations, the effect on the microbiological population can be measured and operations that appear to lower the populations repeated.

A more theoretical but quantitative assessment can be made by comparing the molal concentrations of nutrients to calculate the potential microbiological population. That is, the required ratio of C:N:P changes slightly with each type of microorganism and the nutrient environment they require (Rittmann and McCarty 2001). Microorganisms using organic carbon for growth in an aerobic environment generally require the ratio C144:N16:P1 (Letscher et al. 2015). Lacking organic carbon, the required ratio becomes C88:N16:P1 for autotrophic growth (Rittmann and McCarty 2001).

A ratio of C100:N10:P has been used to generally analyze the potential for microbial growth in water, with C based on carbon that is accessible to microorganisms as a nutrient (Edzwald 2011). See chapter 3 for further discussion.

These ratios can be used to determine which required components are limiting and predict whether excessive microbiological growth could occur. When the concentrations of nutrients in the water approach the nutrient ratios required for various microorganisms' growth, a higher disinfection concentration is needed in the water to counteract the growth.

There should also be awareness that some disinfectants are more powerful than others. Although chloramine disinfection has advantages over free chlorine as a disinfectant in some water systems, a higher dose of chloramine is required to equal the protective level of free chlorine at a lower dose. For instance, it would not be uncommon for a water system using chloramine to use a dose of 2.0 to 2.5 mg/L total chlorine (Zhang et al. 2002) as opposed to 0.3 to 1 mg/L of free chlorine. However, chloramines are able to penetrate biofilms more effectively than free chlorine (Trussell 1999).

Piping configurations. To collect other evidence that MIC might be occurring in a water distribution system, one can start by studying piping configurations that appear to be sources of elevated metals concentrations or metal surface pitting. MIC is typically initiated at the farthest reaches of a distribution system, at dead ends, and at the low points of pipelines. These are the points where water stagnates, sediment accumulates, and disinfection levels are usually low (Bremer et al. 2001). For example, if all pipe failures are occurring underneath the concrete slab foundation of buildings (low points in pipelines), then one would suspect MIC as a possible contributor.

Biofilm analysis. If wet corroded pipe samples can be obtained, the corrosion debris can be analyzed for the presence of microorganisms. An older method involved staining and microscopic techniques (Standridge et al. 2003), the ATP method can estimate the population of microorganisms per area of metal surface or per volume of pipe debris, which is explained further in chapter 3.

Conclusion. Further research concerning the nuances of MIC is greatly needed so that awareness of the role it plays in metal release may be increased. In the meantime, because of the prevalence of microorganisms in drinking water systems, investigators should always consider the possible role microorganisms play in a given water quality problem and prove through quantitative measurements that MIC is either occurring or not (Cantor 2014).

MECHANISMS OF METAL RELEASE BY ACCUMULATION AND TRANSPORT

Another emerging area of concern relates to the accumulation of trace metal deposits (such as those of arsenic, lead, and copper onto iron, manganese, and aluminum) that commonly exist in distribution systems. The physical presence of regulated contaminants on deposit surfaces creates the potential for their uncontrolled chemical and/or physical release into the bulk water (Schock 2005; Maynard and Mast 2006; Schock et al. 2008, 2014; Hill et al. 2010; Wasserstrom et al. 2015). Several case studies have documented contaminant "excursions," whereby contaminant concentrations of copper, arsenic, lead, and radium at the tap far exceed their respective regulated limits, despite extremely low concentrations at system entry points. Contaminant accumulation and release within distribution system scales and sediments are governed by a series of complex and often simultaneously occurring chemical, microbiological, and physical mechanisms, as summarized in Figure 2-5 and discussed more below.

Accumulation Mechanisms

Some researchers describe the accumulation of contaminants in water systems as a constant low-level background flux of particulate inorganic material (Vreeburg et al. 2008, Seth et al. 2004). The material exists in the bulk water because of oxidation of naturally occurring iron and manganese, water treatment carryover, and release from other pipes. These particles may be either large enough to settle under gravitational forces or small enough to remain entrained in the bulk flow. Layers of this particulate matter develop around the entire pipe circumference, not just the pipe invert (Figure 2-6).

As material settles, other chemicals can be sorbed or coprecipitated and become ingrained in the pipe wall scales. As an example of the adsorptive potential of some metals, hydrous manganese oxide is used to adsorb and remove radium as a water treatment process. Manganese oxides also have significant adsorptive capacity for lead (Schock

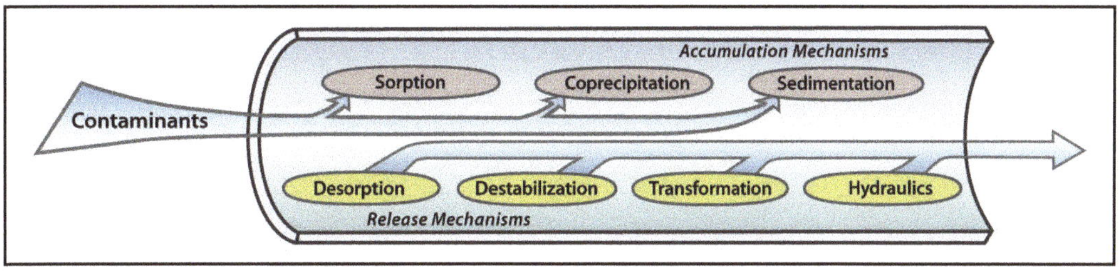

Source: *Hill et al. 2010*

Figure 2-5 Contaminant accumulation and release mechanisms that may occur in distribution system piping

Figure 2-6 Inside a transmission main, showing accumulated material around entire circumference

et al. 2014). In addition, iron-oxide-coated sand or coagulation with iron is used to adsorb and remove arsenic as a water treatment process (Sorg 2002). Aluminum salts are found in soils or used in coagulation and are known to sorb other metals. The focus needs to be on the fact that iron, manganese, and aluminum "substrates" build up in water distribution system pipelines and have similar adsorptive properties (Schock 2005, Friedman et al. 2010).

Studies from one water system, the source water of which included dissolved iron and manganese from the aquifer, showed that iron and manganese scales that precipitate in the distribution system adsorb and accumulate various metals. Visual, mineralogical, and chemical characterization of excavated lead-service-line corrosion scales from that water system revealed that accumulated manganese (and iron) served as high-capacity sinks for lead. Manganese that accumulated from source well water onto pipe scales (up to 10 percent by weight of scale composition) captured and transported lead to consumer taps when the scale was disturbed (Schock et al. 2014). This finding corresponded with lead monitoring studies that had shown lead occurring at the consumers' taps at random times and predominantly in particulate form (Cantor 2006).

Biofilms also play a role in accumulated materials. The particulate inorganic material coming into contact with the pipe wall gets trapped and accumulates via a biofilm-mediated process. Although the mechanisms of attachment and detachment of particles

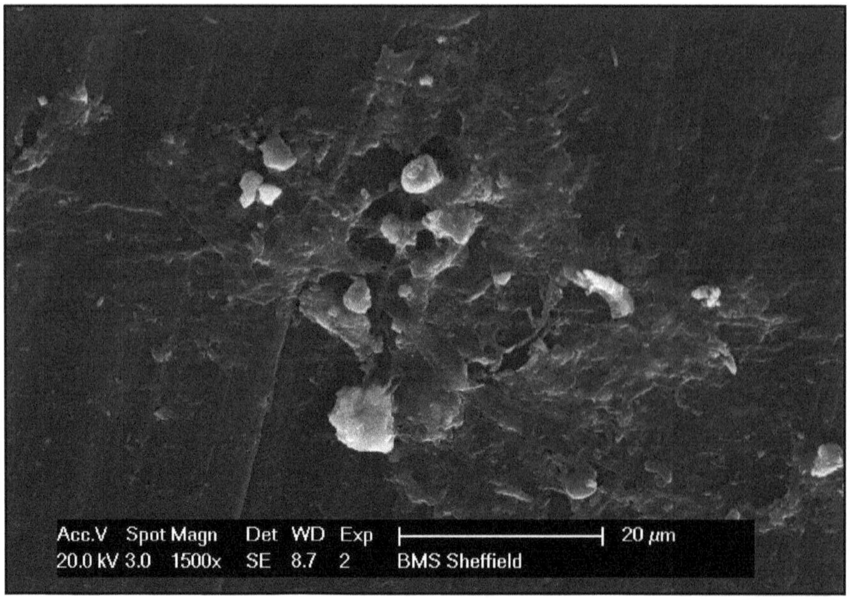

Source: *Speight et al. 2013*

Figure 2-7 Incorporation of inorganic material (which appears white) in biofilm

within a biofilm are not yet fully understood, the accumulation of particles can be considered to be a buildup of cohesive layers of organic and inorganic material attached to the pipe wall (Husband and Boxall 2010). Figure 2-7 shows the incorporation of inorganics into biofilm structure when viewed by scanning electron microscopy.

The resultant accumulation of metals on pipe walls can be many times greater than the initial concentrations entering the distribution system, which is shown in Table 2-3. In addition, significant accumulation and release of iron have been observed in many nonmetal pipes, particularly those downstream of cast-iron pipes (Husband and Boxall 2010, Boxall 2010). Table 2-3 illustrates the mixture of metals that can accumulate within pipes of different materials and how these are not dominated by the local pipe material but rather the cumulative bulk water quality at this point in the distribution system.

Release Mechanisms

Accumulation mechanisms are partially or fully reversible under certain conditions, allowing for particulate or soluble release of accumulated contaminants back into the water supply (Hill et al. 2010). Chemical release can occur from changes in pH and ORP associated with source water blending or source changeovers, treatment process changes, disinfectant residual loss, or biological reactions. These bulk-water chemistry changes can cause desorption of contaminants from iron and manganese substrates or can result in chemical destabilization of existing scales. Water quality conditions that affect distribution system deposit and trace contaminant stability are summarized in Table 2-4 (Friedman et al. 2010).

Physical release can be caused by changes in distribution system hydraulics, such as sudden changes in demand, fire flow events, main breaks, or other changes in flow patterns. Flushing activities, if not conducted properly, can stir up accumulated metals that can be carried to the customers' taps (Vreeburg 2010, Hill et al. 2010, Friedman et al. 2016).

Table 2-3 Example of metal concentrations from flushing operations within one distribution system area comprised of multiple pipe materials

Flush No.	Pipe Material*	Flushing Rate (L/s)
1	PVC	10
2	PVC	5
3	CI	2
4	CI	6
5	PVC	5
6	CI	7
7	CI	8
8	MDPE	8
9	MDPE	8
10	MDPE	11
11	MDPE	11

Flush No.	Turbidity (NTU)†	Iron (μg/L)		Aluminum (μg/L)		Manganese (μg/L)	
		Dissolved	Total	Dissolved	Total	Dissolved	Total
1	4.3	7	11,850	38	1,350	1.5	87.0
2	10.2	9	1,708	20	119	2.3	184.0
3	21.0	17	5,637	23	296	7.9	281.0
4	0.9	7	333	26	47	1.5	23.3
5	3.8	19	2140	22	200	2.4	188.0
6	640.0	26	104,200	39	3,490	14.5	3,670.0
7	3.0	12	707	25	45	1.5	17.5
8	3.7	8	3,553	21	118	1.5	100.0
9	9.0	13	2,056	21	142	1.6	159.0
10	0.5	11	1,229	24	88	1.8	66.7
11	0.8	11	414	23	48	1.5	21.4
Water quality leaving treatment works	<0.15		7.8				<1.5

Flush No.	Zinc (μg/L)		Copper (μg/L)		Lead (μg/L)	
	Dissolved	Total	Dissolved	Total	Dissolved	Total
1	6	71	3	71	0.3	34
2	6	91	3	5	0.3	3.6
3	33	241	3	7	0.3	4.2
4	6	12	3	3	0.3	0.4
5	6	17	3	3	0.3	2.3
6	6	2,640	3	45	0.3	39
7	6	8	3	5	0.3	0.7
8	6	52	3	9	0.3	16.4
9	6	99	3	4	0.3	3.0
10	6	37	3	5	0.3	12.8
11	6	11	3	3	0.3	1.5

*PVC = polyvinyl chloride; CI = cast iron; MDPE = medium-density polyethylene
†NTU = nephelometric turbidity units
Source: *Boxall 2010*

Table 2-4 Water quality conditions that affect deposit and trace contaminant stability

Water Quality Characteristic	Description of Impact
Provide a stable pH within the distribution system (±0.2 units)	The following processes are highly sensitive to pH: adsorption and desorption of trace elements, precipitation/solubilization of minerals capable of serving as accumulation sinks, precipitation/solubilization of trace contaminant compounds, and deposit stability. When implementing purposeful pH adjustment, utilities should be aware of potential release impacts and perform distribution system monitoring.
Provide a stable oxidation–reduction potential (ORP) within the distribution system (±20%)	The nature and stability of mineral deposits is dependent on ORP. These include deposits that may serve as accumulation sinks (e.g., siderite [$FeCO_3$], goethite [α-FeOOH]), as well as chemical precipitates directly involving trace elements (e.g., chromium hydroxide [$Cr(OH)_3$], PbO_2, uraninite [UO_2]).
Provide a stable orthophosphate concentration within the distribution system (±20%)	Orthophosphate can react with common inorganic elements to produce precipitates that may serve as accumulation sinks or low-solubility passivation layers. In either case, it is important to maintain a near-constant concentration to promote stability of these solids. When implementing purposeful phosphate addition, utilities should be aware of potential release impacts and perform distribution system monitoring.
Provide adequate corrosion control	Reduce the formation of iron corrosion scale and tubercles. Reduce the occurrence of red-water episodes. Promote the stability of cement mortar linings. Reduce the leaching of inorganics from cementitious materials.
Avoid uncontrolled blending of surface water and groundwater	Groundwater and surface water supplies typically have very different water quality profiles, including mineral/ionic distribution, NOM concentrations, and ORP. The uncontrolled blending, or periodic switching back and forth, of these different source types can prevent formation of stable corrosion scales and contribute to the release of existing scales and associated contaminants.
Avoid uncontrolled blending of free chlorine and chloramine	The uncontrolled blending, or periodic switching back and forth, can cause dramatic changes in ORP and disinfectant residual concentration, thus impacting scale stability.

Source: *Friedman et al. 2010*

OTHER MECHANISMS OF METAL RELEASE

An investigator should be aware of several other mechanisms of metal release when considering the list of potential causes. The mechanisms discussed in this section are examples of nonuniform corrosion. These mechanisms can be localized in the piping material that is affected and not necessarily pervasive throughout a water system.

Galvanic Corrosion

Metals have individual potentials for giving up electrons. A metal that is more willing to give up an electron is said to be lower in the galvanic series of metals than other metals with less of a tendency. If two dissimilar metals are connected, a flow of electrons can begin like a battery. That is, the metal lower in the galvanic series acts as the anode where electrons are given up. The metal higher in the galvanic series acts as the cathode. The electrons easily flow through both metals just as they would through a wire connecting the anode to the cathode in a battery (Peabody 2001).

In the galvanic series of metals, zinc has more of a tendency to give up electrons than iron. Iron has more of a tendency than lead; and lead, more than brasses and copper

Courtesy of Process Research Solutions, LLC, Madison, Wis.

Figure 2-8 Characteristic horseshoe-shaped pit of erosion corrosion

(Peabody 2001). For example, if a large area of brass is connected to a galvanized iron pipe (zinc coated), then galvanic corrosion will occur, causing pitting of the zinc coating and ultimately the iron pipe underneath (AwwaRF and TZW 1996).

An increased focus on galvanic corrosion has come about because replacement of lead service lines often includes a connection of copper pipe or brass fittings directly to lead pipe downstream. Refer to chapter 5 for further discussion.

Concentration Cells

Early in this chapter, it was pointed out that there are natural microscopic imperfections and variations on a metal surface that cause a flow of electrons to occur between surface locations. Heterogeneity can occur not only in the structure of metal pipe but also in the water in contact with the pipe. For instance, if there is a pocket of higher oxygen concentration in one location versus another, then a corrosion potential can develop (AwwaRF and TZW 1996). For example, low dissolved oxygen in water may be captured between a bolt and a metal surface that is still in contact with higher dissolved oxygen in the general water solution in the main pipe or tank. Or low dissolved oxygen may occur under pipeline debris in contrast with the oxygenated solution outside of the debris (AwwaRF and TZW 1996). These examples are considered to be undergoing corrosion because of *concentration cells.*

Erosion Corrosion

Water at high velocity may physically scour pipe walls, causing surface irregularities and encouraging corrosion to occur locally (AwwaRF and TZW 1996). This scouring typically occurs at pipe bends in systems where turbulence, velocity, and possibly temperature are factors. A characteristic horseshoe-shaped pit is formed, typically just above a pipe bend, in the direction of flow (Figure 2-8).

This type of corrosion is common in hot water recirculating systems in building plumbing where recirculating pump rates can be set too high. The elevated temperature of the water is also a factor in the problem's occurrence.

Stray Electrical Currents and Electrical Grounding

Stray electrical currents and electrical grounding may play a role in metal loss from pipes and pitting of pipes. However, the damage that occurs is found on the exterior of the pipe where the electrical current leaves the pipe and flows into the soil (Duranceau et al. 1998).

This concept is confusing. Some brochures and websites seek to explain lead and copper corrosion to the layperson but do not differentiate between the external pitting phenomenon and internal pitting with increased metal concentration in the water. General statements are made, such as, "Corrosion of pipes is greater if grounding wires are connected to them" (Kissel et al. 2003). There does not appear to be references to show that stray currents or grounding can increase metal concentration in water or increase internal pitting. More research is needed to determine where and when this phenomenon may occur.

Stray electrical currents and grounding currents should, nevertheless, be taken into consideration in a corrosion investigation. If electrical currents are suspected as a possible factor, measurements must be taken to prove or disprove the theory. The phenomenon and measurement techniques are described in an article by Duranceau and colleagues (1998) and in publications by the National Association of Corrosion Engineers (NACE).

Pipe Manufacturing Quality

The manufacturing quality of a pipe can determine the fate of the pipe in an aqueous environment. The more homogeneous a pipe's structure, the less likely that localized anodic and cathodic sites will develop. In copper pipe, for instance, heterogeneity of copper crystal structure can cause pitting to occur when the pipe is in service (AwwaRF and TZW 1996).

Also, residual films, such as carbon films used as lubricant on copper pipe in the manufacturing process, can set up the conditions for pitting (AwwaRF and TZW 1996).

Pipe Installation Practices

Any irregularities on a pipe's surface may set up the conditions for pitting. Spillage of soldering fluxes onto the interior pipe wall is one source of irregularities and connection of dissimilar metals. Incorrect joints and bends may also cause abnormalities. A study on construction practices that can lead to corrosion of metal plumbing is included in Water Research Foundation Project 4164, *Lead and Copper Corrosion Control in New Construction* (Edwards 2011).

System Maintenance

There have also been water systems where the presence of high lead concentrations, specifically particulate lead, has been related to oversights in water system maintenance. In one case, the lining of a building's water tank had deteriorated and formed sludge on the bottom of the tank with debris carried over to the building's water system periodically. The lining contained lead that was carried to the drinking water taps. Gaskets in the system had also deteriorated and lead impurities were carried into the water (Cantor 2006). Therefore, the integrity of tank linings and gaskets may affect water quality.

SUMMARY

This chapter has described the common mechanisms by which metals are released into the water in a distribution system. The most commonly observed phenomena for metals release are

- corrosion by means of carbonate, oxide, and hydroxide solubility of corrosion by-products;
- corrosion by means of chloride and sulfate solubility of corrosion by-products;
- microbiologically influenced corrosion; and
- metals accumulation and transport by chemical scales.

Other possible circumstances include galvanic corrosion, concentration cells, erosion corrosion, pipe manufacturing quality, pipe installation, and system maintenance.

The actual electrochemical process that is termed *corrosion* and is typically thought of as uniform corrosion expressed by carbonate solubility is only one of many aspects involved in metal release.

To understand and control metal release in a water system, one must take a more comprehensive approach to an investigation and measure water quality parameters that describe the above list of possible metal release mechanisms—and more, if the situation warrants. Chapter 3 of this manual explores water quality monitoring and assessment of internal corrosion and increased metals concentrations.

REFERENCES

AWWA (American Water Works Association). 2006. Manual M56. *Fundamentals and Control of Nitrification in Chloraminated Drinking Water Distribution Systems*. Denver, Colo.: AWWA.

AwwaRF and TZW (Awwa Research Foundation and DVGW-Technologiezentrum Wasser). 1996. *Internal Corrosion of Water Distribution Systems*, 2nd ed. Denver, Colo.: AwwaRF.

Boxall, J.B. 2010. Discolouration in water supply: The role of metals. In *COST Action 637 METEAU 4th International Conference on Metals and Related Substances in Drinking Water, Oct. 13–15, Kristianstad, Sweden*. Stockholm: KTH.

Bremer, P.J., B.J. Webster, and D.B. Wells. 2001. Biocorrosion of copper in potable water. *Journal AWWA*, 93(8):82.

Burlingame, G.A., D.A. Lytle, and V.L. Snoeyink. 2006. Why red water? Understand iron release in distribution systems. *Opflow*, 32(12):12.

Cantor, A.F. 2006. Diagnosing corrosion problems through differentiation of metal fractions. *Journal AWWA*, 98(1):117.

Cantor, A.F. 2014. Theoretical lead and copper release into drinking water versus reality. In *Proceedings of the AWWA Water Quality Technology Conference*. Denver, Colo.: AWWA.

Cantor, A.F., D. Denig-Chakroff, R.R. Vela, M.G. Oleinik, and D.L. Lynch. 2000. Use of polyphosphate in corrosion control. *Journal AWWA*, 92(2):95.

Cantor, A.F., J.K. Park, and P. Vaiyavatjamai. 2003. Effect of chlorine on corrosion in drinking water systems. *Journal AWWA*, 95(5):112.

Colling, J.H., P.A.E. Whincup, and C.R. Hayes. 1987. The measurement of plumbosolvency propensity to guide the control of lead in tapwaters. *Journal of the Institution of Water and Environmental Management*, 1(3):263.

Douterelo, I., M. Jackson, and C. Solomon. 2014. How water characteristics and operational conditions can govern the microbial ecology of biofilms in drinking water networks. In *Proceedings of the AWWA Water Quality Technology Conference*. Denver, Colo.: AWWA.

Duranceau, S.J., M.J. Schiff, and G.E.C. Bell. 1998. Electrical grounding, pipe integrity, and shock hazard. *Journal AWWA*, 90(7):40.

Edwards, M. 2011. *Lead and Copper Corrosion Control in New Construction*. Project 4164. Denver, Colo.: Water Research Foundation.

Edwards, M., and N. Sprague. 2001. Organic matter and copper corrosion by-product release: A mechanistic study. *Corrosion Science*, 43:1–18.

Edwards, M., J. Ferguson, and S. Reiber. 1994. The pitting corrosion of copper. *Journal AWWA*, 86(7):74–90.

Edwards, M., M. Schock, and T. Meyer. 1996. Alkalinity, pH, and copper corrosion by-product release. *Journal AWWA*, 88(3):81–94.

Edzwald, J.K. 2011. *Water Quality and Treatment: A Handbook on Drinking Water*, 6th ed. Denver, Colo.: AWWA.

Ehrlich, L.H., and D.K. Newman. 2009. *Geomicrobiology*, 5th ed. Boca Raton, Fla.: Taylor and Francis.

Escobar, I.C., and A.A. Randall. 2002. Case study: Ozonation and distribution system biostability. *Journal AWWA*, 93(10):77.

Fallowfield, H.J., M. Critchley, N. Bolton, and N. Cromar. 2011. Microbiologically Influenced Corrosion of Copper and Galvanized Steel Pipe. Presented at Australasian Corrosion Association Symposium on Microbiologically Influenced Corrosion, Melbourne, Aug. 10 and 11, 2011.

Federal Register. 1991. National Primary Drinking Water Regulations; Lead and Copper; Final Rule. *Federal Register*, 56(110):26460.

Fleming, K.K., G.W. Harrington, and D.R. Noguera. 2005. Nitrification potential curves: A new strategy for nitrification prevention. *Journal AWWA*, 97(8):90.

Friedman, M.J., A.S. Hill, S.H. Reiber, R.L. Valentine, G. Larsen, A. Young, G.V. Korshin, and C.Y. Peng. 2010. *Assessment of Inorganics Accumulation in Drinking Water System Scales and Sediments*. Denver, Colo.: Water Research Foundation.

Friedman, M., A. Hill, S. Booth, M. Hallett, L. McNeill, J. McLean, D. Stevens, D. Sorensen, T. Hammer, W. Kent, M. De Haan, K. MacArthur, and K. Mitchell. 2016. *Metals Accumulation and Release Within the Distribution System: Evaluation and Mitigation*. Denver, Colo.: Water Research Foundation.

Goldscheider, N., D. Hunkeler, and P. Rossi. 2006. Review: Microbial biocenoses in pristine aquifers and an assessment of investigative methods. *Hydrogeology Journal*, 14:926–941.

Gomez, C., R. Hozalski, and T. LaPara. 2014. Characterization of drinking water distribution system biofilm communities using next-generation illumina sequencing of 16S RRNA. In *Proceedings of the AWWA Water Quality Technology Conference*. Denver, Colo.: AWWA.

Grasha, L.A., W.W. Risner, and M.K. Davis. 1981. Update on corrosion study at the Metropolitan Water District of Southern California. *Awwa Research Foundation Water Quality Research News*, 24(9):2–3.

Hidmi, L., and Edwards, M. 1999. Role of temperature and pH in $Cu(OH)_2$ solubility. *Environmental Science and Technology*, 33:2607–2610.

Hill, A.S., M.J. Friedman, S.H. Reiber, G.V. Korshin, and R.L. Valentine. 2010. Behavior of trace inorganic contaminants in drinking water distribution systems. *Journal AWWA*, 102(7):107.

Holm, T.R., and M.R. Schock. 1991. Potential effects of polyphosphate products on lead solubility in plumbing systems. *Journal AWWA*, 83(7):76.

Husband, P.S., and J.B. Boxall. 2010. Field studies of discolouration in water distribution systems: Model verification and practical implications. *Journal of Environmental Engineering*, 136, 86–94.

Ives, D., and A. Rawson. 1962. Electrochemical theory of general corrosion. *Journal of the Electrochemical Society*, 109(6).

Jaisi, D.P., R.K. Kukkadapu, L.M. Stout, T. Varga, and R.E. Blake. 2011. Biotic and abiotic pathways of phosphorus cycling in minerals and sediments: Insights from oxygen isotope ratios in phosphate. *Environmental Science and Technology*, 45(15):6254–6261.

Ji, P., J. Parks, M. Edwards, and A. Pruden 2014. Microbiome composition of standardized premise plumbing rigs: Comparing effect of pipe material and water chemistry. In *Proceedings of the AWWA Water Quality Technology Conference*. Denver, Colo.: AWWA.

Kissel, D.E., P.R. Vendrell, and J.H. Atiles. 2003. *Your Household Water Quality: Lead and Copper*. Consumer information brochure from the University of Georgia Cooperative Extension Services.

Klueh, K.G., and R.B. Robinson. 1988. Sequestration of iron in groundwater by polyphosphates. *Journal of Environmental Engineering*, 114(5):1192.

Knobeloch, L., C. Schubert, J. Hayes, J. Clark, C. Fitzgerald, and A. Fraundorff. 1998. Gastrointestinal upsets and new copper plumbing—Is there a connection? *Wisconsin Medical Journal* (Jan.).

Korshin, G., S. Perry, and J. Ferguson. 1996. Influence of NOM on copper corrosion. *Journal AWWA*, 88(7):36–47.

Koudleka, M., J. Sanchez, and J. Augustynski. 1982. On the nature of surface films formed on iron in aggressive and inhibiting polyphosphate solutions. *Journal of the Electrochemical Society*, 129(6):1186.

Larson, T.E., and R.V. Skold. 1958. *Laboratory Studies Relating Mineral Quality of Water to Corrosion of Steel and Cast Iron*. Circular 71. Champaign, Ill.: Illinois State Water Survey. pp. 43–46.

Letscher, R.T., J.K. Moore, Y.-C. Teng, and F. Primeau. 2015. Variable C:N:P stoichiometry of dissolved organic matter cycling in the Community Earth System Model. *Biogeosciences*, 12(1):209–221.

Little, B., P. Wagner, and F. Mansfield 1992. An overview of microbiologically influenced corrosion. *Electrochimica Acta*, 37(12):2185–2194.

Lytle, D.A., and M.R. Schock. 2005. Formation of Pb(IV) oxides in chlorinated water. *Journal AWWA*, 97(11):102.

Madigan, M.T., and J.M Martinko. 2006. *Brock Biology of Microorganisms*, 11th ed. Upper Saddle River, N.J.: Prentice Hall.

Maynard, B., and D. Mast. 2006. Composition of interior scales on lead source materials. In *Proceedings of the AWWA Water Quality Technology Conference*. Denver, Colo.: AWWA.

Nguyen, C.K., B.N. Clark, K.R. Stone, and M.A. Edwards. 2011. Role of chloride, sulfate, and alkalinity on galvanic lead corrosion. *Corrosion*, 67(6):065005-1–065005-9.

Peabody, A.W. 2001. *Peabody's Control of Pipeline Corrosion*, 2nd ed. Houston: NACE International.

Perry, R.H., and D. Green (eds.). 1984. *Perry's Chemical Engineers' Handbook*, 6th ed. New York: McGraw-Hill.

Raskin, L., A. Pinto, D. Marcus, and G. Dick 2014. Metagenomic insights into bacteria that dominate drinking water microbial communities. In *Proceedings of the AWWA Water Quality Technology Conference*. Denver, Colo.: AWWA.

Rego, C.A., and M.R. Schock. 2007. Discovery of unforeseen lead level optimization issues for high pH and low DIC conditions. In *Proceedings of the AWWA Water Quality Technology Conference*. Denver, Colo.: AWWA.

Rittmann, B.E., and P.L. McCarty 2001. *Environmental Biotechnology: Principles and Applications*. New York: McGraw-Hill.

Rodgers, M. 2014. Impact of corrosion control on publically owned treatment works. In *Proceedings of the AWWA Water Quality Technology Conference*. Denver, Colo.: AWWA.

Rogers, J.R., and P.C. Bennett. 2004. Mineral stimulation of subsurface microorganisms: Release of limiting nutrients from silicates. *Chemical Geology*, 203:91–108.

Sarver, E., and M. Edwards. 2012. Inhibition of copper pitting corrosion in aggressive potable waters. *International Journal of Corrosion*. Article ID 857823. doi:10.1155/2012/857823

Schneider, O.D., J. Parks, M. Edwards, A. Atassi, and A. Kashyap. 2011. *Comparison of Zinc Versus Non-Zinc Corrosion Control for Lead and Copper*. Report 4103. Denver, Colo.: Water Research Foundation.

Schock, M.R. 1980. Response of lead solubility to dissolved carbonate in drinking water. *Journal AWWA*, 72(12):695.

Schock, M.R. 2005. Distribution systems as reservoirs and reactors for inorganic contaminants. In *Distribution System Water Quality Challenges in the 21st Century—A Strategic Guide*. Edited by M.J. McPhee. Denver, Colo.: AwwaRF.

Schock, M.R., and D.A. Lytle. 2014. Demystifying Copper Corrosion. Presented at New England Water Works Association Conference.

Schock, M.R., and A.M. Sandvig. 2006. Long-term impacts of orthophosphate treatment on copper levels. In *Proceedings of the AWWA Annual Conference and Exposition*. Denver, Colo.: AWWA.

Schock, M., D. Lytle, and J. Clement. 1995. *Effect of pH, DIC, Orthophosphate and Sulfate on Drinking Water Cuprosolvency*. Cincinnati, Ohio: US Environmental Protection Agency, Office of Research and Development, National Risk Research Laboratory.

Schock, M.R., D.A. Lytle, A.M. Sandvig, J. Clement, and S.M. Harmon. 2005. Replacing polyphosphate with silicate to solve lead, copper and source water iron problems. *Journal AWWA*, 97(11):84.

Schock, M.R., R.N. Hyland, and M.M. Welch. 2008. Occurrence of contaminant accumulation in lead pipe scales from domestic drinking-water distribution systems. *Environmental Science and Technology*, 42(12):4285–4291.

Schock, M.R., A.F. Cantor, S. Triantafyllidou, M.K. Desantis, and K.G. Scheckel. 2014. Importance of pipe deposits to Lead and Copper Rule compliance. *Journal AWWA*, 106(7):E336.

Seth, A., R. Bachmann, J. Boxall, A.J. Saul, and R. Edyvean. 2004. Characterisation of materials causing discolouration in potable water systems. *Water Science and Technology*, 49, 27–32.

Sorg, T.J. 2002. Iron treatment for arsenic removal neglected. *Opflow*, 28(11):15.

Speight, V., I. Douterelo, R. Sharpe, K. Fish, and J. Boxall. 2013. Impact of hydraulic events on organic and inorganic material in distribution systems. In *Proceedings of the AWWA Water Quality Technology Conference*. Denver, Colo.: AWWA.

Srinivasan, S., and G. Harrington. 2007. Biostability analysis for drinking water distribution systems. *Water Research*, 41(4):2127.

Standridge, J., B. Hoffman, and L. Peterson. 2003. *Brown Deer Pipe Corrosion Study*. Madison, Wis.: Wisconsin State Laboratory of Hygiene.

Trussell, R.R. 1999. Safeguarding distribution system integrity. *Journal AWWA*, 91(1):46–54.

Turek, N., L. Kasten, D. Lytle, and M. Goltz. 2011. Impact of plumbing age on copper levels in drinking water. *Journal of Water Supply*, 60(1):1–15.

USEPA (US Environmental Protection Agency). 2016. *Optimal Corrosion Control Treatment Evaluation Technical Recommendations for Primacy Agencies and Public Water Systems*. EPA 816-B-16-003. Washington, D.C.: USEPA.

Van der Kooij, D. 1992. Assimilable organic carbon as an indicator of bacterial regrowth. *Journal AWWA*, 84(2):57.

Volk, C.J., and M.W. LeChevallier. 2000. Assessing biodegradable organic matter. *Journal AWWA*, 92(5):64.

Vreeburg, J. 2010. *Discolouration in Drinking Water Systems: The Role of Particles Clarified*. London: IWA Publishing.

Vreeburg, J.H.G., D. Schippers, J.Q. Verberk, and J.C. Van Dijk. 2008. Impact of particles on sediment accumulation in a drinking water distribution system. *Water Research*, 42, 4233–4242.

Wang, Y., Y. Xie, and D.E. Giammar. 2012. *Lead(IV) Oxide Formation and Stability in Drinking Water Distribution Systems*. Report 4211. Denver, Colo.: Water Research Foundation.

Wasserstrom, L., M. Schock, and S. Miller. 2015. Practical Implications From Observed Lead Pipe Scale Mineralogy in a Blended Phosphate Treated System (slides). Presented at AWWA Annual Conference and Exposition, Anaheim, Calif., June 7–10.

Wolfe-Simon, F., J.S. Blum, T.R. Kulp, G.W. Gordon, S.E. Hoeft, J. Pett-Ridge, J.F. Stolz, S.M. Webb, P.K. Weber, P.C.W. Davies, A.D. Anbar, and R.S. Oremland. 2011. A bacterium that can grow by using arsenic instead of phosphorus. *Science*, 332:1163–1166.

Zhang, M., M.J. Semmens, D. Schuler, and R.M. Hozalski. 2002. Biostability and microbiological quality in a chloraminated distribution system. *Journal AWWA*, 94(9):112.

Zinder, B., J. Hertz, and H.R. Oswald. 1984. Kinetic studies on the hydrolysis of sodium tripolyphosphate in sterile solution. *Water Research*, 18(5):509.

Chapter 3

Water Quality Monitoring and Assessment of Internal Corrosion and Increased Metals Concentrations

Melinda Friedman, Confluence Engineering Group, LLC
Christopher P. Hill, Arcadis, Inc.
Richard Giani, CH2M
Andrew Hill, Confluence Engineering Group, LLC
Andrew Jacque, Water Quality Investigations, LLC
Matthew McFadden, M3P Services, LLC

INTRODUCTION

This chapter is divided into two main sections: Part 1 elaborates on the fundamentals of internal corrosion outlined in chapter 2. It emphasizes the critical water quality parameters that affect internal corrosion and that should be considered when trying to understand the causes of corrosion and metals release issues. Part 2 discusses water quality monitoring programs focused on characterizing corrosion problems; presents a general approach to assessing the cause of corrosion-related events; and discusses different approaches for organizing, monitoring, and planning for investigative diagnosis of corrosion and metals release problems. Table 3-1 summarizes many of the key points discussed in this chapter.

PART 1: WATER QUALITY PARAMETERS AFFECTING CORROSION AND METALS RELEASE

The Carbonate Balance

Chapter 2 mentioned that the nature of various metallic carbonate species influence development and characteristics of protective coatings on the interior of distribution system piping. Therefore, a thorough understanding of the factors that influence the presence of carbonate and bicarbonate species in water, called the *carbonate balance*, is necessary in developing an effective corrosion control program.

The pH value. The pH value of water is a measure of the concentration of the hydrogen ion concentration (H^+ or H_3O^+) (Edzwald and Tobiason 2011). The pH scale ranges from 0 to 14—values less than 7.0 are considered acidic, values greater than 7.0 are considered alkaline or basic, and 7.0 is considered neutral (Economic and Engineering Services and Illinois State Water Survey 1990). Drinking water pH values typically range from 6.0 to 10. At higher pH values, there is less of a tendency for metal surfaces in contact with drinking water to dissolve and dissociate, which is why pH adjustment is a common component of an effective corrosion control treatment strategy. Maintaining a consistent target pH throughout the distribution system is always critical to minimizing lead and copper levels at the tap and minimizing discolored-water complaints, even if other corrosion control methods are employed.

Another important consideration with regard to pH is its impact on other water quality parameters discussed in this chapter. It plays a significant role in the carbonate balance in that it impacts buffer capacity and dissolved inorganic carbon (DIC) concentrations. However, it also influences other corrosion-related parameters, such as oxidation–reduction potential (ORP) and corrosion inhibitor effectiveness. These impacts are discussed at greater length throughout this chapter.

Alkalinity. Alkalinity is the capacity of water to neutralize acid (Stumm and Morgan 1996). It is the sum of carbonate (CO_3^{2-}), bicarbonate (HCO_3^-), and hydroxide (OH^-) anions and is typically reported as milligrams per liter as calcium carbonate (mg/L as $CaCO_3$). Waters with high alkalinities tend to have high buffering capacities or a strong ability to resist changes in pH. Low-alkalinity waters are less able to neutralize acids or resist changes in pH.

Buffer intensity or buffer capacity. Buffer intensity is a measure of the resistance of water to upward or downward changes in pH and is a function of pH, DIC, and alkalinity (USEPA 2016). Bicarbonate and carbonate ions are the most important buffering species in most drinking water supplies. At pH ≥9.0, silicate ions also provide some buffering. Buffering intensity from carbonate species is normally greatest at approximately pH 6.3 and above 9.0, and lowest in the range of pH 8.0 to 8.5 (see Figure 3-1).

Buffer intensity can have a significant impact on calcium carbonate precipitation. Stumm (1960) demonstrated that hard waters with high buffer intensities and slightly positive Langelier saturation index (LSI) values deposited less calcium carbonate than soft waters with slightly negative LSI. Similar results have been observed by others (Sontheimer et al. 1979). It was also found that corrosion scales with more than 5 to 10 percent $CaCO_3$ (by weight) may be less protective than those with lower $CaCO_3$ concentrations (Snoeyink and Wagner 1996).

Table 3-1 Chapter 3 key points

Water Quality Impacts on Corrosion
• Understanding the carbonate balance and its impact on corrosion is critical in that it affects passivation, which is the ability of mineral scales to inhibit the electrochemical process of corrosion. • pH is a key component of an effective corrosion control program. Metal dissociation is less likely to occur at higher pH values (above neutral or higher, depending on water chemistry). • Maintaining a consistent pH throughout the distribution system is critical to maintaining effective corrosion control. • pH, alkalinity, and dissolved inorganic carbon are parameters that define the carbonate balance. • Buffering intensity from carbonate species is highest at pH 6.3 and above 9.0, and lowest in the 8.0-to-8.5 range. • Oxidants, and more specifically the residual disinfectant in the distribution system, impact the finished water oxidation–reduction potential (ORP). Changes in the disinfectant that change ORP have the potential to significantly impact corrosion and metals release in the distribution system and home plumbing. • Anions such as chloride, sulfide, and sulfate can change the corrosive characteristics of the water with respect to different metals. • The presence of sulfide may significantly increase corrosion rates and the occurrence of pinhole leaks in copper plumbing. • Iron and manganese can serve as accumulators of lead and other regulated metals on various piping and home plumbing materials, and can contribute to ongoing particulate and soluble releases of lead and trace inorganics. • Chloride is a critical anion in copper corrosion and pitting. • The chloride-to-sulfate mass ratio can have a significant impact on the corrosion rate of lead from soldered joints and brass. • Microbiologically influenced corrosion (MIC) is often overlooked as a cause of increased metals release. Maintaining biostability is essential to an effective corrosion control program. • MIC will be more prevalent in water systems where the source water has sufficient concentrations of organic carbon, organic nitrogen, and phosphate to support microbial growth. • Corrosion "inhibitors" can be effective for controlling corrosion and metals release; however, it is critical that the mechanism by which an inhibitor works is understood. • Water quality may not always be the cause of increased metals release; hydraulic and other factors may result in increased metals concentrations at residential and commercial user end points.
Water Quality Monitoring Programs
• Lead and copper compliance monitoring may not be sufficient for determining and maintaining effective corrosion control treatment. More frequent monitoring of lead, copper, and additional parameters is recommended. • It is necessary to establish a water quality baseline when developing a corrosion control program to evaluate impacts of changes in water quality on corrosion. • A monitoring program should include routine monitoring at the source, entry to the distribution system, and various locations in the distribution system, including at the tap. • The number of monitoring locations, the frequency of monitoring, and the number of parameters monitored will be dependent on source water quality, changes in treatment procedures, distribution system piping materials, type of disinfectant used, the presence of lead service lines, and other variables including economic factors.
Assessment of Corrosion-Related Water Quality Problems
• A convenient way to classify corrosion for the purpose of diagnosis is (1) chemically influenced, (2) microbiologically influenced, and (3) physically influenced. • Identifying the cause of a corrosion-related problem requires an evaluation of how the above-mentioned three categories of corrosion are contributing to that problem. • Lack of an adequate monitoring program will make it difficult to determine the cause of corrosion-related water quality problems.

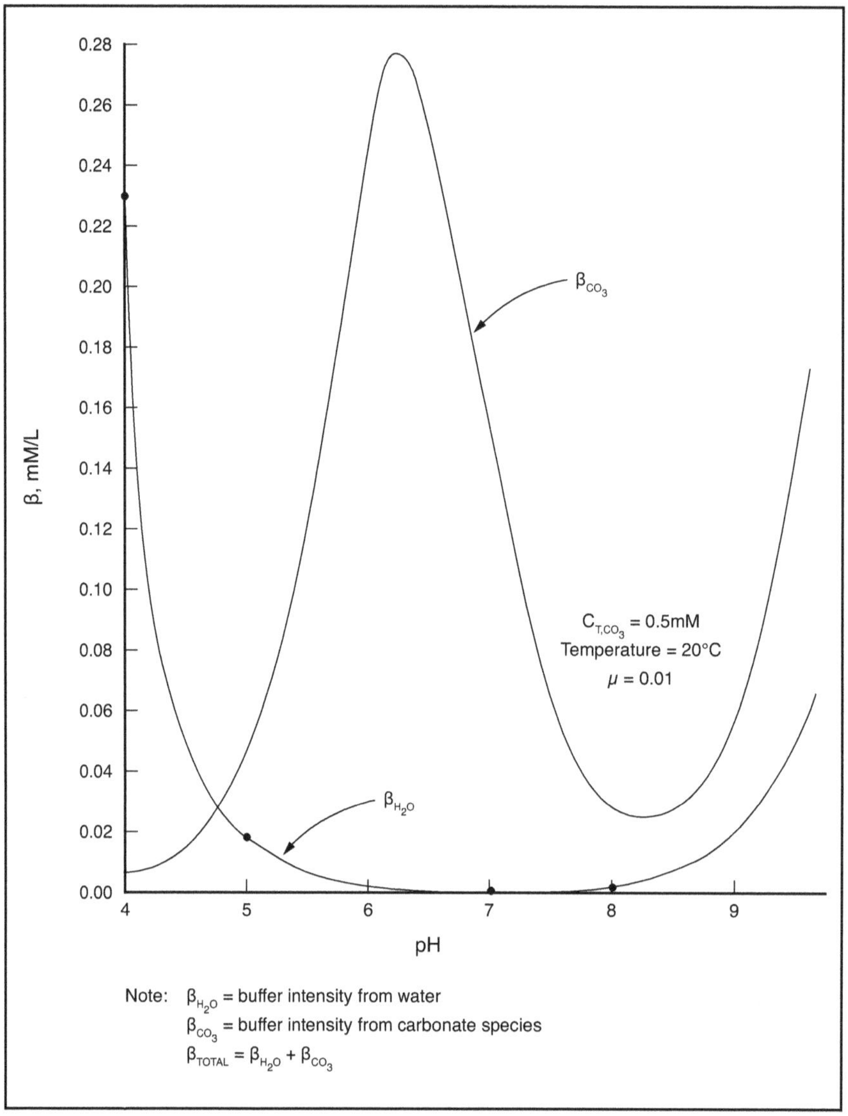

Source: *Snoeyink and Wagner 1996*

Figure 3-1 Variation of buffer intensity with pH

Corrosion reactions require an anode (electron donor) and a cathode (electron receptor). In drinking water distribution systems, these reactions at the metal surface result in pH shifts at the anode and cathode that further contribute to the corrosion reaction (Snoeyink and Wagner 1996). Stumm (1960) showed that iron corrosion rates decreased in a uniform manner as the buffer intensity increased. It was theorized that this pattern was because waters with higher buffer intensity experienced less of a pH increase at the cathode and less of a pH decrease at the anode. Other studies have found similar results (Clement and Schock 1998, Van Der Merwe 1988, Pisigan and Singley 1987).

Total dissolved solids/ionic strength. Total dissolved solids (TDS) can have an impact on corrosion, though the effect may be less than other corrosion-related parameters. High TDS concentrations are generally associated with high concentrations of ions (e.g., Na^+, Ca^{2+}, Mg^{2+}, Cl^-, CO_3^{2-}, SO_4^{2-}) that increase the conductivity of the water. As discussed in chapter 2, corrosion is an electrochemical reaction in which electrons from the anodic surface are transferred to the cathodic surface. The increased conductivity resulting from high TDS concentrations increases the ability of the water to complete the electrochemical circuit and conduct a corrosive current.

If sulfate and chloride are major anionic contributors to the TDS, the TDS is likely to show increased corrosivity toward iron-based materials. If the TDS is composed primarily of bicarbonate and hardness ions, the water may not be corrosive toward iron-based or cementitious materials but may be highly corrosive toward copper (Schock 1999). Low-TDS waters may also be corrosive and increase lead solubility. Low-TDS waters often have a strong tendency to dissolve (corrode) materials with which they are in contact in an attempt to reach electroneutrality. Recall from chapter 2 that uniform corrosion is an electrochemical process in which the water solution in contact with the cathode provides the chemicals to accept the electrons donated by the pipe wall. The pipe wall, acting as an anode, then releases the oxidized metal ion to the water. Water has a limited capacity to accept dissolved species. Thus, although high-TDS waters have many electron receptors, low-TDS waters have the ability to accept a large number of anions and cations, resulting in a subsequent dissolution of existing pipe scales and corrosion of pipe surfaces.

TDS is a surrogate for the ionic strength of a solution. The ionic strength is a measure of the force of the electrostatic field caused by the presence of ions in a solution. More simply, the presence of anions and cations in solution increases the conductivity of the solution and can increase corrosion unless offset by passivating layers on the pipe surface (Schock 1999). Ionic strength can be determined as follows (Economic and Engineering Services and Illinois State Water Survey 1990):

$$I = 2.5 \times 10^{-5} \times TDS$$

Where:

I = ionic strength

TDS = total dissolved solids concentration, in mg/L

Dissolved inorganic carbon. DIC is the sum of all DIC-containing species and is one of the most critical parameters for controlling internal corrosion. It includes dissolved aqueous carbon dioxide gas (CO_2 or H_2CO_3), bicarbonate ion (HCO_3^-), and carbonate ion (CO_3^{2-}) in a particular water, and DIC is usually expressed as milligrams per liter of carbon (mg/L as C) or milligrams per liter of calcium carbonate (mg/L as $CaCO_3$). Although DIC and alkalinity are similar, they are not the same water quality parameter.

DIC varies according to water temperature, pH, ionic strength, and alkalinity. An example of this relationship is provided in Figure 3-2. DIC also significantly impacts the buffer intensity of water. Figure 3-3 shows that as the DIC concentration increases, the buffer capacity of the water also increases. Because DIC controls the buffer capacity in most water systems, sufficient DIC is required to maintain a stable pH throughout the distribution system for control of copper and lead (Schock and Lytle 1995), especially at pH values less than 8.3. Above pH 8.3, larger increases in buffering capacity can generally be achieved through pH increases, as compared to DIC addition (Schock et al. 2014a).

In the presence of bicarbonate, copper corrosion undergoes a critical transition somewhere between pH 7.0 and 8.5 (Edwards et al. 1994b). Scale formed at pH 7.0 catalyzes oxygen reduction and increases the overall corrosion rate. At pH values between 7.5 and 8.5, DIC complexes dominate copper speciation, resulting in increased copper solubility. This effect is most prominent in the DIC range of 1 to 20 mg/L as C. Finally, at pH 8.5 and above, oxygen reduction rates are unchanged, and copper dissolution is inhibited, resulting in a passivation of the copper surface.

Hardness. Hardness is a characteristic that represents the presence of dissolved multi-valence cations, primarily calcium and magnesium, in water and is reported as an equivalent concentration of calcium carbonate ($CaCO_3$) (Post et al. 2011). When sufficient

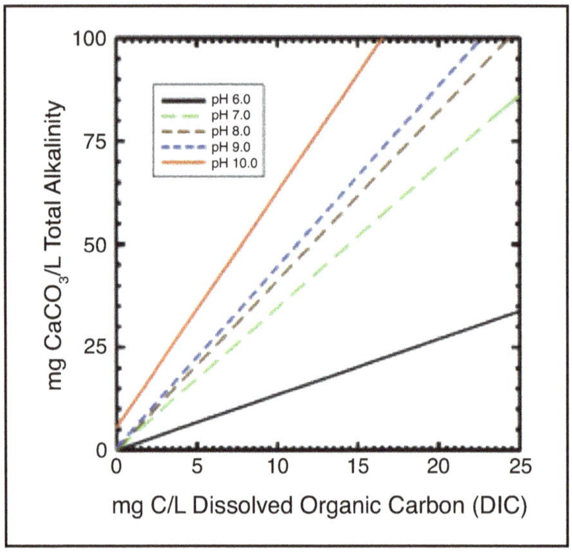

Source: *Schock et al. 2014a*

Figure 3-2 Relationship between alkalinity and DIC for various pH levels (pH = 6.0 to 10.0, I = 0.005, T = 25°C)

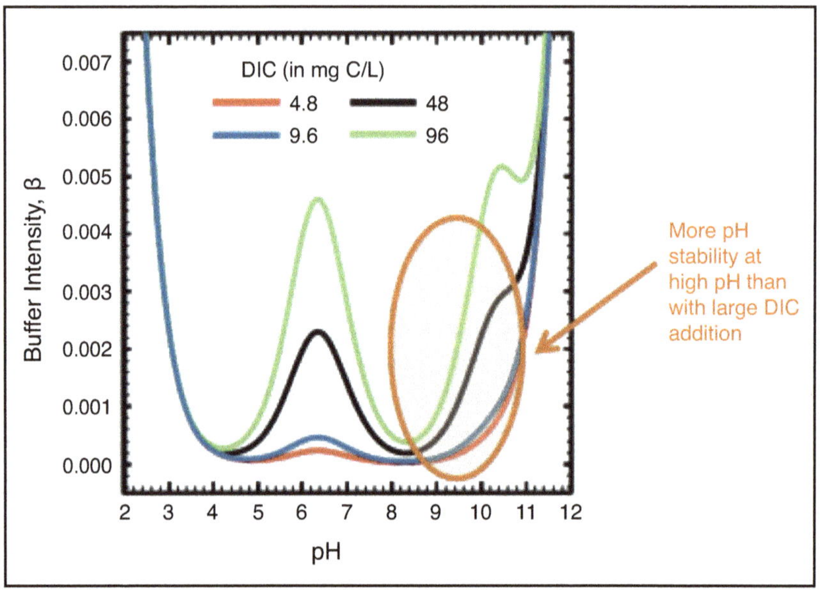

Source: *Schock et al. 2014a*

Figure 3-3 Effect of DIC (measured as C) on buffer intensity

calcium and alkalinity are present in waters with pH greater than the saturation pH, a scale of $CaCO_3$ may form on the inside of distribution piping. However, the formation of calcium carbonate scale can interfere with corrosion control when other methods, such as phosphate passivation, are employed (Schock and Lytle 2011), and calcium carbonate scales may be porous and/or poorly adhered to the pipe wall, allowing continued transport of metal ions into water. Therefore, calcium carbonate precipitation is not considered an effective form of corrosion control (USEPA 2016).

Hardness must be taken into consideration when corrosion control is selected and implemented because it can create scaling problems within the treatment plant, distribution system infrastructure, and customer plumbing. In this regard, hardness is an important parameter to be considered in developing a corrosion control program and evaluating the amount of pH adjustment that can be made without causing scale problems. However, it is not a stand-alone indicator of the corrosive nature of a water. Trying to control corrosion by setting pH as high as possible without considering the potential for scaling can be problematic.

Corrosivity

Corrosion is the deterioration of a substance or its properties because of an electrochemical reaction with its environment (Schock 1999). Although this term may occasionally be used to describe the deterioration of cement mortar lining and other cement-based materials used in piping systems, in the context of this manual, *corrosivity* refers to the oxidation and subsequent dissolution of the metal surfaces of the interior of pipe walls. From a regulatory perspective, the concern with lead and copper corrosion is the dissolution, or increased solubility, of lead and copper. From an aesthetic perspective, the oxidation of ferrous iron to ferric iron has the potential to create red water and other problems.

The most significant water quality parameters with regard to the corrosivity of water are pH, alkalinity, and DIC. Balancing pH, alkalinity, and DIC concentrations enables the formation of carbonate passivating layers. Similarly, pH values less than the prescribed corrosion inhibitor ranges can cause a decrease in chemical passivation and thus cause an increase in metal solubility.

Edwards and colleagues (1994b) demonstrated that increasing the pH from 5.5 to 7.0 reduced corrosion rates by half in new (unaged) copper pipe. Stone et al. (1987) demonstrated that increasing the pH from 6.0 to 8.0 reduced the copper corrosion rate by half in both new and aged copper pipe; however, the increase in pH had no effect on zinc (galvanized iron) pipe. In a review of the first round of Lead and Copper Rule (LCR) monitoring data, it was observed that no copper action level (AL) exceedances were reported for systems having a pH greater than approximately 8.0 (Schock and Lytle 1995).

In a survey of nearly 400 US drinking water utilities, Dodrill and Edwards (1995) observed that for utilities not adding a phosphate inhibitor at pH less than 8.4, lead release was significantly lower when alkalinity was 30–74 mg/L as $CaCO_3$ compared to alkalinity less than 30 mg/L. With regard to copper release, it was reported that problems meeting the copper AL occurred under two water quality scenarios: (1) pH less than 7.0 and alkalinity less than 30 mg/L and (2) pH less than 7.8 and alkalinity greater than 90 mg/L. In a review of the results of the first round of LCR monitoring data, Schock and Lytle (1995) noted that copper exceedances tended to be highest at low alkalinities (<25 mg/L as $CaCO_3$) and increasingly greater over 75 mg/L as $CaCO_3$.

A direct correlation between red-water incidents and finished water alkalinity has also been observed. Horsley et al. (1998) noted that the frequency of red-water occurrence increased significantly when the finished water alkalinity dropped below 60 mg/L. In this case, the authors concluded that the optimum finished water alkalinity is likely to vary with finished water chemistry and system configuration (e.g., the types of pipe present and other potential corrosion concerns, such as lead). However, alkalinity adjustment should be considered when attempting to address red-water problems. Reiber (2006) observed that iron release more than doubled following a decrease in alkalinity from approximately 30 mg/L to 10 mg/L in a pilot distribution system at a pH of approximately 7.5. Imran et al. (2005) determined that decreases in alkalinity concentrations resulted in increases in red-water occurrence in pilot distribution systems in which calcite ($CaCO_3$) formation was used as a primary means of corrosion control.

Portland Water District (PWD), Maine, experienced firsthand the significance of DIC and the impact on buffer intensity with regard to lead corrosion control (Boissonneault 1994). In the early 1990s, like much of the United States, PWD was struggling to comply with the LCR. At the time, PWD used a low-alkalinity (<5 mg/L as $CaCO_3$), low-DIC (<2 mg/L as C) source water that had a poor buffering capacity and was extremely corrosive. Corrosion control at the time included pH adjustment to approximately 7.6 and addition of a zinc orthophosphate for the control of red water. Multiple corrosion control strategies—including pH adjustment, pH adjustment plus zinc orthophosphate, and pH adjustment plus zinc polyphosphate—were tried. None of these strategies were effective for the control of lead corrosion. It was ultimately determined that the preferred lead control strategy was increasing the DIC using carbon dioxide or sodium bicarbonate to increase the buffer intensity and to aid in maintaining a stable finished water pH, combined with zinc orthophosphate addition and a target finished water pH of 7.6 (Boissonneault 1994).

It has been suggested that (cation exchange) softening is likely to create a more corrosive water because of the removal of calcium and subsequent increase in TDS that results from its use. However, it has been observed that cation exchange softening has little effect on lead, copper, and galvanized iron corrosion rates (Sorg et al. 1999) because the ion exchange process does not significantly alter the parameters that have the most significant impact on corrosion: pH, alkalinity, and, most importantly, DIC. This observation demonstrates the importance of DIC in selecting, implementing, and maintaining an effective corrosion control strategy. However, certain treatment technologies can increase lead in drinking water by changing water quality parameters that impact lead release.

In the anion exchange process used for removal of contaminants such as nitrate and uranium, freshly regenerated ion exchange resin removes bicarbonate ions, causing reduction in pH and total alkalinity during the initial 100 bed volumes (BVs) of a run. Raising the pH of the treated water may be required at the beginning of a run (100–400 BVs) to avoid corrosion (Clifford 1999, Wang et al. 2010, Clifford et al. 2011). Similarly, frequent regeneration of an anion exchange resin can have an impact on corrosion. In a case study in Maine, frequent regeneration of the ion exchange resin was instituted to reduce the levels of uranium in the waste stream (residuals). This resulted in a significant and continual decrease of pH and subsequent leaching of copper and lead into the drinking water (Lowry 2009, 2010). Since reverse osmosis completely removes alkalinity in water, it will continually lower treated water pH and increase its corrosivity. Therefore, the product water pH must be adjusted to avoid corrosion issues in the distribution system, such as the leaching of lead and copper (Schock and Lytle 2011, USEPA 2012).

Oxidants

The presence of oxidants such as dissolved oxygen, chlorine, and chloramine can significantly impact internal corrosion. Many instances have been reported where failure to consider the presence or absence of a particular oxidant and its potential impact on corrosion has resulted in lead and copper exceedances or increases in red-water complaints. As discussed below, it is generally the change in oxidant condition itself that can result in increased metals release from existing pipe scales and sediments.

Dissolved oxygen. Dissolved oxygen (DO) can play important roles in both corrosion reactions and metals release. As described in chapter 2, oxygen can serve as an electron acceptor in the corrosion cell, allowing for metal oxidation at the pipe surface and release of ionized metal species into the water. Thus, new metal surfaces that are exposed to water containing DO will corrode faster compared to anaerobic water. Schock et al. (1998) noted that copper pipe in water containing oxidizing agents will continue to corrode until either all of the oxygen is depleted or precipitated oxide films arrest the rate of corrosion. Schock and Lytle (2011) indicate that high DO levels may facilitate the

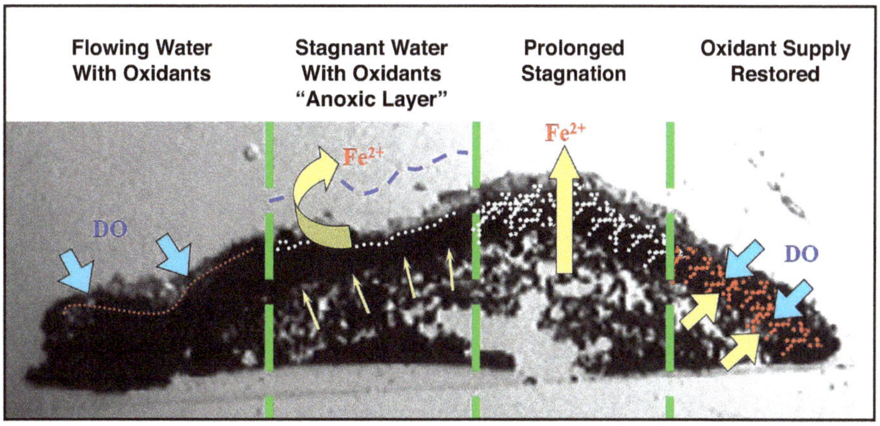

Source: *Sarin et al. 2000*

Figure 3-4 Impact of oxidants on iron release from corrosion scales

production of different, more protective mineral oxide and hydroxide solid corrosion products compared to environments with limited oxygen availability.

With regard to metals release from existing corrosion scales, DO has been reported to both increase and decrease red-water episodes. Imran et al. (2005) identified DO as a contributing factor to red-water episodes in a pilot distribution system based on the results of nonlinear regression modeling predictions of changes in apparent color. In these studies, increases in DO concentrations resulted in increases in red-water occurrence. Conversely, other studies have demonstrated the importance of increasing DO concentrations under stagnant flow conditions to reduce soluble iron release from corroded iron pipes (Sarin et al. 2003). As shown in Figure 3-4, in flowing water with oxidants present, Fe^{3+} (ferric iron) is present at the surface of tubercles and is relatively impermeable. Under stagnant conditions, loss of DO (and disinfectant residual) can create an anoxic layer that increases the porosity of tubercles, allowing for the release of soluble Fe^{2+} (ferrous iron) from the porous tubercles and/or the conversion of surficial Fe^{3+} to soluble Fe^{2+}. Once the oxidant supply is restored, the pores in the outer layer will be filled back in, reducing further soluble iron diffusion from inside the tubercle.

In another example, a utility in the Pacific Northwest recently converted from oxygen-saturated surface water to low-oxygen (≤1 mg/L) groundwater. In low-flow areas of the distribution system supplied by unlined cast-iron pipe, severe discoloration/scale destabilization occurred. Pipe loop studies using heavily tuberculated 6-in. unlined cast-iron pipes extracted from the distribution system demonstrated that increasing DO levels to the range of 5 to 8 mg/L reduced both the rate and the magnitude of iron and color development (Figures 3-5 and 3-6) after 48 hours of stagnation (Friedman and Blain 2014).

As concluded by Schock and Lytle (2011), the interactions of oxygen with pipe surfaces and their effects on the physical properties and mineralogy of scales still need substantial research. However, based on current knowledge, increased DO from advanced oxidation processes would not be expected to cause corrosion problems in distribution systems that have historically practiced chemical disinfection.

Oxidation–reduction potential. ORP, also frequently referred to as *redox potential*, is a measure of water's capability to oxidize and is reported as electrical potential (volts, V, or millivolts, mV). Disinfectants are, by their very nature, oxidants, and as such, adding a disinfectant will increase ORP in the water. Figure 3-7 shows the effect of oxidant type (free chlorine, monochloramine, chlorine dioxide, potassium permanganate, and dissolved oxygen) on redox potential in weakly buffered deionized water (DIC = 5 mg/L as C).

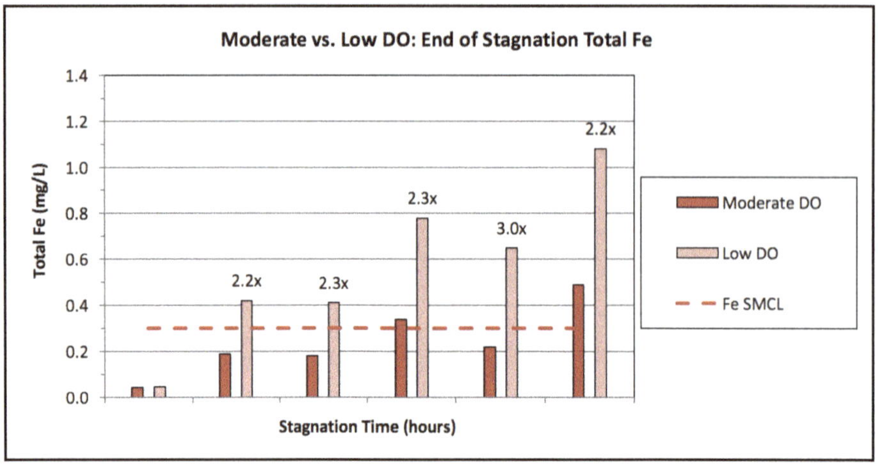

DO = dissolved oxygen; SMCL = secondary maximum contaminant level

Source: *Friedman and Blain 2014*

Figure 3-5 Comparison of iron levels with and without DO during pipe rig stagnation trials

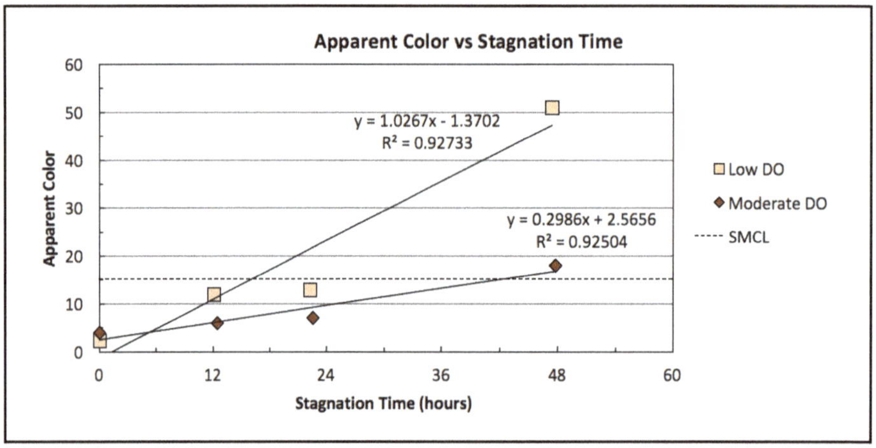

DO = dissolved oxygen; SMCL = secondary maximum contaminant level

Source: *Friedman and Blain 2014*

Figure 3-6 Comparison of color concentration and rate of color development with and without DO during pipe rig stagnation trials

Some disinfectants are stronger oxidants than others. For example, oxygen can quickly oxidize ferrous (soluble) iron but is very slow to oxidize soluble manganese. Chlorine will nearly instantaneously oxidize soluble iron, but it could take several minutes or hours to oxidize soluble manganese. Chloramine, which has recently become a more common disinfectant used to reduce chlorinated disinfection by-products, will oxidize ferrous iron but not as readily as chlorine. Therefore, utilities that switch from chlorine to chloramine will reduce ORP values in the distribution system. Similarly, as disinfectant residual is lost within the distribution system, ORP levels can be reduced significantly, especially in the absence of DO.

The Washington Aqueduct utility, the wholesale purveyor to Washington, D.C., switched from using free chlorine for secondary disinfection to chloramine in November 2000, which appeared to have had some role in the increased lead concentrations found

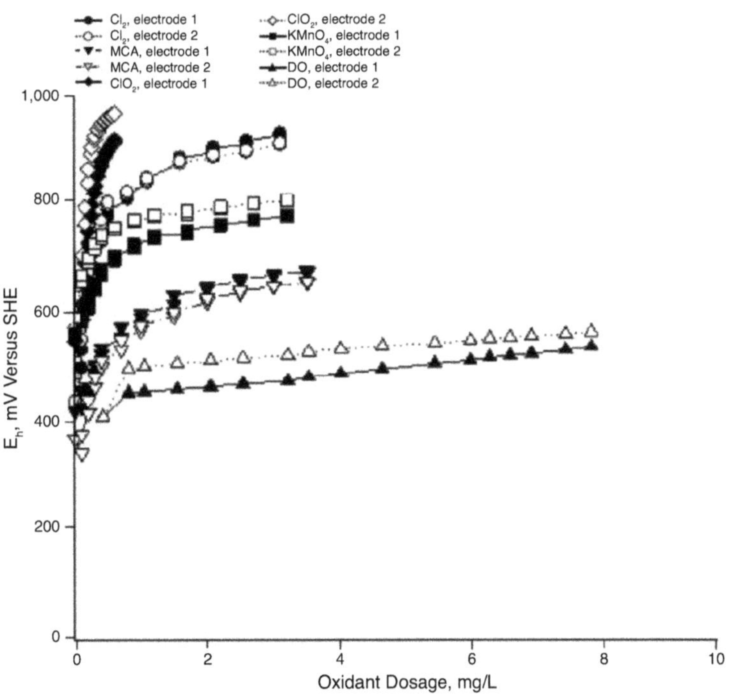

Cl$_2$—chlorine, ClO$_2$—chlorine dioxide, DO—dissolved oxygen,
Eh—half-reaction electrode potential (measured in millivolts),
KMnO$_4$—potassium permanganate, MCA—monochloramine,
SHE—standard hydrogen electrode

Dissolved inorganic carbon = 5 mg/L C; pH = 8; temperature = 23°C

Source: *Copeland and Lytle 2014*

Figure 3-7 Effect of oxidant type on ORP in weakly buffered deionized water (DIC = 5 mg/L as C)

within the system. Bench-scale experiments (Edwards and Dudi 2004) seemed to confirm that a switch from free chlorine to chloramine can accelerate the dissolution of lead under conditions similar to those in Washington, D.C., particularly where brass and bronze in-line devices (faucets, fixtures, water meters, etc.) are present. Lytle and Schock (2005) conducted long-term lead precipitation experiments with chlorinated water to determine the water quality conditions and pathways leading to the formation of lead dioxide (PbO_2) scales and the stability of PbO_2 scales in response to changes in ORP. Although some have implicated chloramine as the cause of the elevated lead levels in Washington, D.C., experts thus far have suspected that free chlorine actually served as a form of corrosion control treatment (Reiber 2004, Lytle and Schock 2005). The hypothesis is that it was not chloramine addition per se that was contributing to lead leaching but rather the absence of the free chlorine previously employed. Other researchers have shown that chloramine also altered the oxidation state of existing lead scales, which resulted in a shift in mineral species from Pb^{4+} to Pb^{2+} (Schock and Giani 2005, Renner 2006, Reiber 2004, Lin and Valentine 2008). Prior to chloramine conversion, lead scales in the Washington, D.C., system had primarily consisted of less-soluble Pb^{4+} species. However, the change in redox potential caused an equilibrium shift to more-soluble Pb^{2+}, which resulted in an increase in tap lead concentrations. Note that higher ORP can be more effective for corrosion control because higher ORP generally results in the formation of more stable metal complexes during passivation (USEPA 2007).

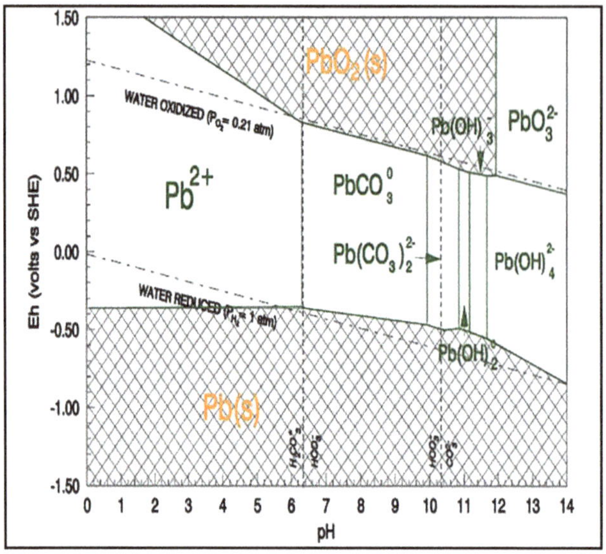

SHE = standard hydrogen electrode

Adapted from Schock and Lytle 2011

Figure 3-8 Lead Pourbaix diagram for 0.015 mg/L Pb, 18 mg C/L DIC, and I = 0

The typical drinking water ORP ranges from about 400 to 800 mV, depending on the type and concentration of disinfectant used. Higher ORP levels will increase oxidation rates in drinking water. Schock and Giani (2005) were able to demonstrate that scale formation on pipe walls may change with changes in ORP. Friedman and Blain (2014) demonstrated the impacts of disinfectant residual loss, associated ORP drop, and increased iron and manganese release from tuberculated unlined cast-iron pipe under stagnation conditions. Large decreases in ORP (>100 mV, especially below approximately 200 mV) may cause existing scales to begin to weaken, dissolve, or change chemical form (e.g., Fe^{3+} to Fe^{2+}, or Pb^{4+} to Pb^{2+}). This behavior includes iron oxide or lead oxide scales that might have formed on existing lead pipes because of high ORP values.

Pourbaix diagrams show the impact of electrochemical potential (or ORP) and pH on metal speciation and mineral formation and are frequently referred to as potential–pH diagrams, Eh–pH diagrams, or pE–pH diagrams. Figures 3-8, 3-9, and 3-10 present typical Pourbaix diagrams for a lead–water system, an iron–water system, and a copper–water system, respectively. Similar diagrams are available for these and other metals and drinking water constituents in a variety of water environments from numerous sources in the printed and electronic literature (e.g., Stumm and Morgan 1996, Snoeyink and Jenkins 1980). Several sources provide electronic tools for constructing Pourbaix diagrams (e.g., PhreePlot [phreeplot.org], Pymatgen [pymatgen.org], and Chesta [www.aqua.mtl.kyoto-u.ac.jp/chestaEng.html]). Pourbaix diagrams do not predict corrosivity or metal solubility but can provide insight into metals speciation and mineral stability. Using Figure 3-8, for example, if the potential of a chlorinated water is 1.0 V and the pH is 8.0, it can be predicted that the predominant lead species present will be PbO_2 solid (Pb^{4+}), which is likely to be the most stable precipitate for maintaining a lead scale on distribution pipes. However, if the Eh and/or pH drops (because of a change in residual type/dose or other microbial reactions, for example), the predominant lead species present will be $PbCO_3$ (Pb^{2+}), which, as discussed previously, is a significantly more soluble species of lead.

Oxidant type and residual. In addition to the ORP impacts associated with chlorine and chloramines described previously, these secondary disinfectants can affect corrosivity, metal release, or both by impacting the corrosion potential of the surface of the metal or direct reaction mechanisms (Schock and Lytle 2011). Free chlorine residuals can impact

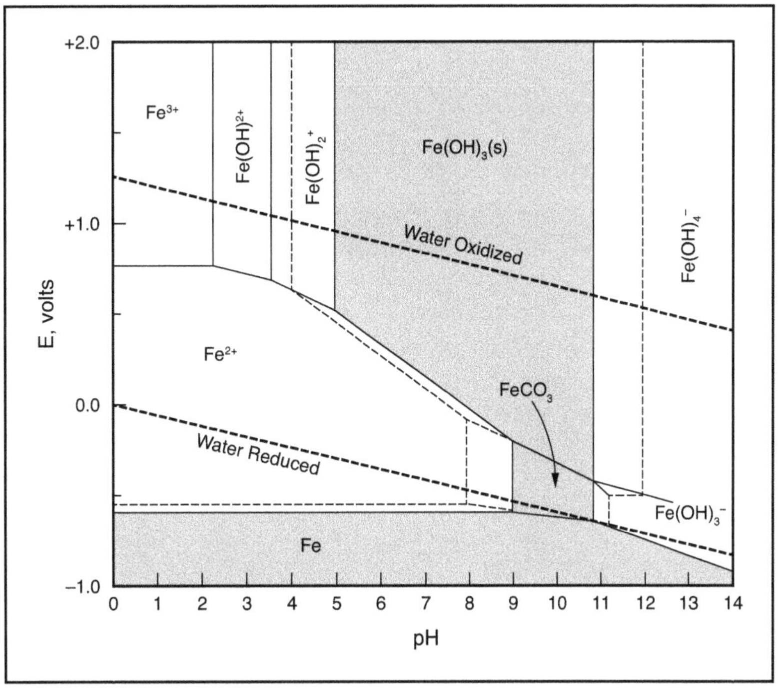

Note: Stability fields are shown for dissolved iron species activities of 0.1 mg/L (— solid lines) and 1.0 mg/L (– – – dashed lines). Dissolved carbonate species concentrations are 4.8 mg/L as C (4×10^{-4} M).

Source: *Schock 1999*

Figure 3-9 Iron Pourbaix diagram for a carbonate-containing water at 25°C and I = 0

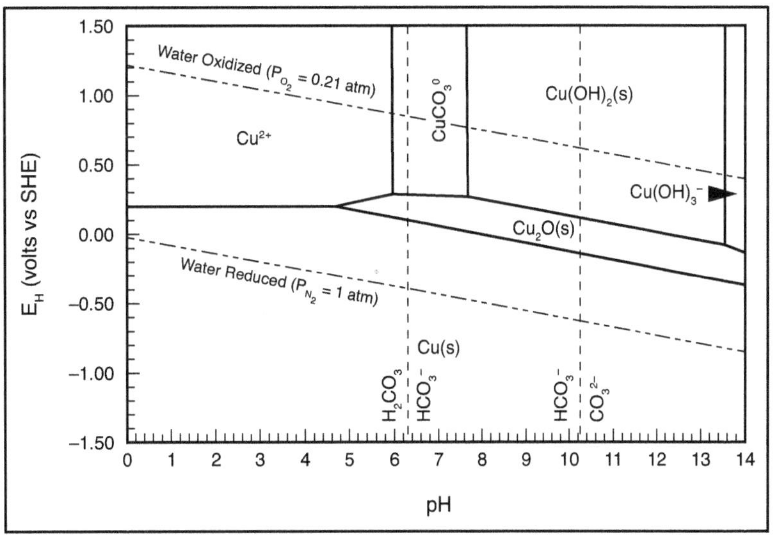

SHE = standard hydrogen electrode

Source: *Schock et al. 1995*

Figure 3-10 Copper Pourbaix diagram

metal oxidation rates and the nature of scales that form on the interior of distribution system piping. In poorly buffered waters, addition of free chlorine in the form of chlorine gas can lower the pH, thereby increasing the corrosion potential of the surface metal, whereas the addition of free chlorine in the form of hypochlorite can raise the pH.

It is often *the change* in disinfectant residual that can result in impacts to metals release, such as conversion of more-stable Pb^{4+} to less-stable Pb^{2+} because of ORP changes. The initial introduction of chlorine can increase corrosion rates, as observed by Pisigan and Singley (1987). Corrosion rates of both mild steel and copper were found to be higher in the presence of free chlorine than in its absence. It was further observed that the corrosion rates of aerated water were less than those of non-aerated water containing free chlorine, suggesting that free chlorine is more corrosive than DO. Aerated water containing free chlorine was the most corrosive of the conditions evaluated. Stone et al. (1987) observed that free chlorine residual concentration had a significant impact on both aged and fresh (new) copper pipe corrosion but did not impact zinc (galvanized iron) corrosion.

Anions

The presence of some anions (e.g., chloride, sulfide, sulfate, nitrate) can play a significant role in the internal corrosion of distribution system and home plumbing materials. This section briefly discusses some of the more important anions with regard to corrosion and their potential impacts on corrosion.

Sulfide. The presence of sulfide can significantly increase corrosion rates and the occurrence of pinhole leaks in copper plumbing. Jacobs et al. (1998) demonstrated that sulfide-containing waters were substantially more aggressive with regard to copper. Even following the elimination of sulfide from the source water and removal of previous pipe scales, pipes previously exposed to sulfide corroded at substantially higher rates than pipes never exposed to sulfide. Sulfide is also a by-product of reduction of sulfate by sulfate-reducing bacteria (SRB) and can be an indicator of microbiologically influenced corrosion (MIC) (Jacobs et al. 1998).

Chloride. Chloride has been demonstrated to be a critical anion in the initiation of copper corrosion and pitting (Ives and Rawson 1962). However, Edwards et al. (1994b) found that copper surfaces passivated in the presence of chloride at pH ≥ 7.0 in aged piping systems.

Imran et al. (2005) determined that increases in chloride concentrations in the range of 22 to 70 mg/L resulted in increases of red-water occurrence in pilot distribution systems. Lytle et al. (2003) observed that total iron release increased by nearly 50 percent in the presence of 100 mg/L of chloride. Addition of orthophosphate at a concentration of 3 mg/L as PO_4 was able to reduce total iron release to normal levels. Following discontinuation of orthophosphate addition, total iron release again increased but not significantly, suggesting some iron–phosphate mechanism that produced lasting benefits (Lytle et al. 2003).

Sulfate. Imran et al. (2005) determined that increases in sulfate concentrations resulted in increases in red-water occurrence in pilot distribution systems. Edwards et al. (1994b) demonstrated that sulfate was the most significant of the anions studied (including carbonate, chloride, perchlorate, and nitrate) with regard to copper corrosion rates. In these studies, water with high sulfate concentrations (100 mg/L) exhibited increased corrosion rates with aging.

Chloride-to-sulfate ratio. The chloride-to-sulfate mass ratio (CSMR $Cl^-:SO_4^{2-}$) has a significant impact on the leaching of lead from soldered joints and brass in some cases (Triantafyllidou and Edwards 2007) and in lead plumbing (Edwards et al. 1999, Nguyen et al. 2010). Nguyen et al. (2010) describe the significance of CSMR for predicting lead leaching caused by galvanic corrosion, indicating that this is due to the net effect of relative concentrations of chloride and sulfate at the lead anode surface. Chloride is related to increased

Table 3-2 Impact of coagulant changes on lead corrosion

	Primary Coagulant	
Parameter	Alum	Ferric Chloride
pH*	7.5	7.4
Alkalinity, mg/L as $CaCO_3$*	75.0	70.0
Sulfate, mg/L*	74.0	34.0
Chloride, mg/L*	24.0	38.0
Chloride–sulfate ratio	0.32	1.11
Total organic carbon, mg/L	2.5	1.9
Total trihalomethanes, μg/L	65.0	40.0
Number of distribution samples >15 μg/L Pb	0/71	12/71

*Finished water.

Source: *Carlson et al. 2000*

lead release and sulfate to reduced lead release. Changes in coagulant to improve natural organic matter removal can significantly alter the CSMR in some waters. Edwards et al. (1999) reported that in a utility survey, 100 percent of the utilities with $Cl^-:SO_4^{2-}$ ratios less than 0.58 met the 0.015 mg/L Pb action level. However, of those facilities with $Cl^-:SO_4^{2-}$ ratios greater than 0.58, only 36 percent met the AL. Nguyen et al. (2010) found similar results, with strong impacts caused by increasing CSMR to 0.4–0.6 but significant lead leaching not occurring where the CSMR was already high.

Carlson et al. (2000) noted that one utility switching from alum to ferric chloride to enhance total organic carbon (TOC) removal and reduce distribution system trihalomethane concentrations resulted in that utility exceeding the lead AL. The change in coagulants resulted in an increase in the $Cl^-:SO_4^{2-}$ ratio; however, all other key corrosion parameters remained relatively constant. Table 3-2 summarizes the impact of coagulant changes on system Pb levels for this utility.

The city of Columbus (Ohio) Division of Water (DOW) experienced similar detrimental effects on lead corrosion following a switch from alum to ferric chloride. Figure 3-11 shows the $Cl^-:SO_4^{2-}$ ratio and 90th percentile lead concentration for the DOW service area. A "system average" $Cl^-:SO_4^{2-}$ ratio is presented because DOW operates three plants—two surface water and one groundwater. The surface water plants were switched over from alum to ferric chloride in July 1995 and September 1996. The groundwater plant is purely a lime-softening plant and does not use a coagulant (Hill 2016).

The issues faced by Stafford County (Virginia) Department of Utilities are similar to those of many utilities and exemplify the difficulties of complying with multiple simultaneous drinking water regulations and, in this case, wastewater treatment requirements. The county operates two conventional surface water treatment plants. Prior to 2001, both of the plants used alum for coagulation and zinc orthophosphate for corrosion control. In 2001, concerns about zinc concentrations at the county wastewater treatment plants motivated a change to an orthophosphate–polyphosphate blend for corrosion control. In 2003, when the Stage 1 Disinfectants and Disinfection Byproducts Rule required additional TOC removal at one of the plants, the decision was made to switch from alum to ferric chloride at that treatment plant. After the switches to the orthophosphate–polyphosphate blend and ferric chloride, 4 of 32 tap samples taken in 2003 had lead concentrations exceeding the 15 μg/L AL. In 2004, 12 of 60 samples had lead concentrations exceeding the AL. The county has since changed from the orthophosphate–polyphosphate blend to straight orthophosphate and planned to convert from ferric chloride to ferric sulfate (Triantafyllidou 2006).

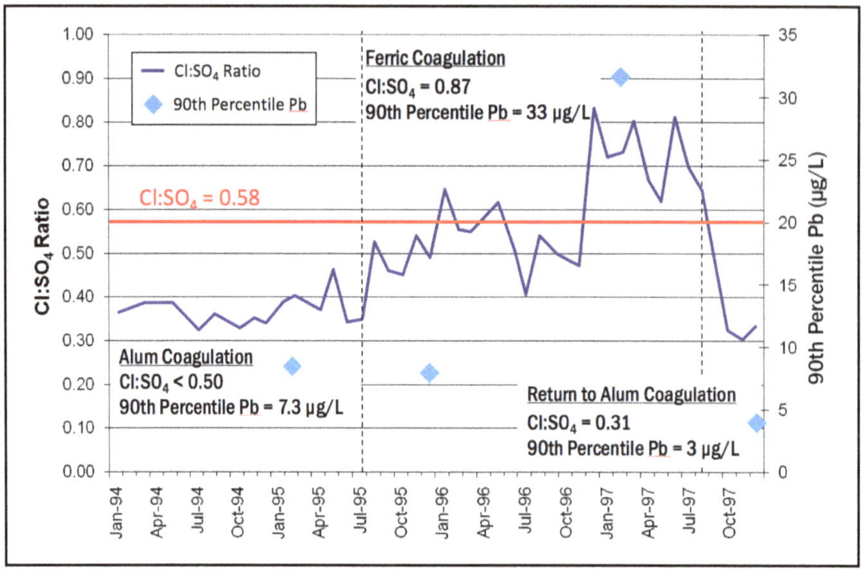

Source: *Hill 2016*

Figure 3-11 Impact of chloride-to-sulfate ratio on lead corrosion

Although it is unclear which treatment change had the most significant impact on the increased lead corrosion (the change to polyphosphate or switch to ferric chloride), both factors have previously been identified as responsible for increases in lead corrosion (Carlson et al. 2000, Edwards and McNeill 2002). Edwards and McNeill (2002) noted that polyphosphate can be effective for the control of particulate iron and manganese in finished water but is generally not recommended for lead and copper corrosion control. After conversion to ferric chloride, Stafford County observed an increase in the $Cl^-:SO_4^{2-}$ ratio from 0.29 to 0.38 prior to the switch, to 4.75 after the switch. The change from ferric chloride to ferric sulfate reduced the $Cl^-:SO_4^{2-}$ ratio to well below the levels present prior to 2003 and reduced 90th percentile lead concentrations within 1 to 2 months (Triantafyllidou and Edwards 2007).

Other anions. Edwards et al. (1994b) found that copper surfaces activated (i.e., increased corrosion rates with aging) in the presence of nitrate, perchlorate, and sulfate over the pH range of 5.5 to 10. It was suggested that the relative aggressiveness of various anions toward copper after aging was

$$HCO_3^- > SO_4^{2-} > NO_3^- > ClO_4^- > Cl^- \text{ (at pH 7.0)}$$

$$SO_4^{2-} \geq ClO_4^- > NO_3^- > HCO_3^- > Cl^- \text{ (at pH 8.5)}$$

Biostability

The biostability of a water, or its propensity to support microbiological activity, can affect corrosion and metals release in a number of ways, as discussed in chapter 2. For example, microbiologically influenced corrosion, or MIC, of copper pipe may be the result of (1) the production of acid metabolites on the metal surface, (2) binding of copper by bacterial metabolites (extracellular polysaccharides) that inhibits the natural bactericidal effects of copper, and (3) alteration of the normal cupric oxide film from the incorporation of bacterial cells and polymers (Bremer et al. 2001). MIC could also represent a mechanism to acquire phosphorus bound in copper pipe, similar to the strategy used by microbes to acquire phosphorus from aquifer materials in a phosphorus-limited environment (Bennett et al. 2001, Rogers and Bennett 2004, Wu et al. 2008).

Several factors can contribute to MIC of copper pipe, including (Bremer et al. 2001)

- use of soft water with low pH (neutral or neutral-alkaline);
- high suspended solids and assimilable organic carbon content;
- long-term or periodic stagnation in the pipe, which produces widely fluctuating oxygen concentrations;
- dead ends or long horizontal runs that are susceptible to sediment accumulation;
- low or nonexistent levels of chlorine;
- maintenance of water temperatures that promote rapid growth and activity of naturally occurring bacteria that form biofilms on the pipe wall; and
- lack of an adequate monitoring program to periodically evaluate water quality and pipe wall conditions.

The type and relative concentration of nutrients present, and environmental conditions, will select for a set of bacteria that can grow. Assessment of the stoichiometric ratio of dissolved organic carbon to total nitrogen (ammonia, nitrate–nitrite, and total Kjehldahl nitrogen) to total phosphorus in conjunction with the presence of DO, sulfate, and alkalinity can help identify nutrient limitations and the potential for biofilm to form for nutrient fixation (nitrogen and/or carbon), and/or corrosion to acquire phosphorus, as discussed in chapter 2. Specific parameters that should be considered when assessing impacts of microbial growth and biostability on corrosion and metals release are summarized in the sections that follow.

Assimilable organic carbon (AOC). AOC is the fraction of biodegradable organic carbon consisting of small organic compounds that can be used easily by microorganisms as a food source. AOC typically enters the distribution system at the source but can also be added to water by autotrophic bacteria that fix carbon dioxide to make organic carbon or fermenting bacteria in biofilm that reduce complex organics to simpler forms. Generally, AOC makes up only 0.1 to 9.0 percent of the TOC present in a water (Escobar and Randall 2001). Van der Kooij (1992) reported that the AOC level must be less than 10 μg/L to prevent microbiological activity in the absence of a disinfectant.

The use of strong oxidants (such as ozonation, chlorine dioxide, or high dosages of free chlorine) may break down larger organic molecules into smaller molecules, resulting in an increase in AOC concentrations entering the distribution system. For example, failure to follow ozonation with biologically active filters can result in significant increases in distribution system AOC concentrations, which can increase microbial activity within the distribution system. It has been theorized that in at least one instance, this increase in AOC resulted in an increase in the occurrence of MIC and pinhole leaks in copper pipes (Cantor et al. 2003). Subsequent evaluation in the same system determined that increased disinfectant residuals eliminated copper pipe failures and appeared to have confirmed the occurrence of MIC (Cantor et al. 2006).

Disinfectant residual. Reiber et al. (1987) observed that corrosion rates varied significantly with fluctuations in distribution system free-chlorine residual concentrations in both aged and new copper pipes in a low-alkalinity (15–17 mg/L as $CaCO_3$) water. Bremer et al. (2001) noted that MIC of copper was predominant in pipes with low or no disinfectant present. Zhang et al. (2002) observed that maintaining a total chlorine (i.e., chloramine) residual greater than 2 mg/L was essential for maintaining biostability and controlling microbial regrowth. It was also noted that even in waters with moderate to high AOC concentrations, biostability could be maintained given proper maintenance of a disinfectant residual. Such factors are critical to the prevention of MIC of copper pipe.

Other bacterial nutrients. Ammonia, sulfate, nitrate, and phosphorus can serve as potential nutrient sources for a variety of bacteria commonly found in distribution and home plumbing systems.

Ammonia is a nutrient source for ammonia-oxidizing bacteria (AOB) and is the precursor to nitrification. Ammonia typically enters the distribution system from the source water, but it can also be introduced by the breakdown of chloramine and biological matter (amino acids) or the fixation of nitrogen gas by nitrogen-fixing bacteria (diazotrophs) to create ammonia. AOB oxidize ammonia to form nitrite (partial nitrification), which in turn may serve as a nutrient source for nitrite-oxidizing bacteria (NOB). NOB oxidize nitrite to form nitrate (complete nitrification). The nitrification process itself consumes alkalinity and may result in localized pH depression, increasing the potential for corrosion.

Sulfate is a nutrient source for sulfate-reducing bacteria (SRB) and is a key factor in MIC of copper piping. Further, in oxygen-deficient conditions, SRB reduce sulfate to sulfide, which is a critical anion in the pitting corrosion of copper pipe (Jacobs et al. 1998).

Similarly, phosphorus is a required nutrient for microbial growth; it is a component in nearly all microbial cell membranes and is required for production of adenosine triphosphate, the basic energy carrier in nearly all living cells. Phosphorus enters drinking water from the source water, is added at the source for corrosion control and/or iron/manganese sequestration, or is added by MIC of copper pipe (Bennett et al. 2001, Rogers and Bennett 2004, Wu et al. 2008). The presence of phosphorus may increase microbiological activity in distribution and home plumbing systems. Increased microbiological activity increases the potential for MIC.

Heterotrophic plate count (HPC). Heterotrophic bacteria utilize organic carbon present in distribution systems and household plumbing as a source of energy. The HPC test provides another measure of microbial activity in water and thus an indirect measure of nutrient availability, but it too has the same limitations that are associated with the AOC test. The HPC test is commonly used as an indicator of microbiological water quality in the industry, but it is important to note that heterotrophs represent only a small fraction of microbes in a water sample. There are two common HPC methods: the standard plate count agar and R2A agar method (Standard Method 9215). It is recommended that the R2A agar method be used for all HPC analyses because it is more sensitive and is likely to result in a more accurate depiction of microbiological activity (Smith 2006).

Elevated HPCs compared to baseline concentrations indicate that increased microbiological activity is likely to occur in areas of the distribution system or home plumbing system in which there have been significant losses in disinfectant residual, where AOC levels are sufficient to encourage microbial regrowth, or in which nitrification is occurring. Cantor et al. (2003) observed that MIC was occurring in areas of a distribution system with AOC concentrations between approximately 1.4 and 126 μg/L in which the total chlorine (i.e., monochloramine) residual concentration was negligible. In addition, areas of chloraminated distribution systems where no residual is present are known to be more susceptible to nitrification (Fleming et al. 2005).

Adenosine triphosphate (ATP). Nearly every living cell uses ATP as its basic energy carrier (Wolfe-Simon et al. 2011). The relatively new second-generation ATP test provides a method to assess total microbial activity in a water sample, with the ability to test microbial activity regardless of electron donor and/or electron acceptor utilized. An estimate of microbial occurrence can be achieved by using an average concentration of 1 femtogram ATP/microbe (microbial equivalent) based on testing with *Escherichia coli*. Results can be obtained within 15 minutes of sample collection compared to two or three days for the HPC test. The ATP test assesses total microbial occurrence compared to the HPC test, which detects about 1% of total microbial occurrence. ATP detection in water is covered by ASTM D4012 (last revised in 2015) and reflects current advancements in testing technology. Although this is closer to a true measure of microbial occurrence, it may underestimate the population in nutrient-deprived conditions when internal ATP stores could be low or overestimated if macroorganisms are present in the sample analyzed. Nevertheless, the ATP test allows an investigator to identify locations of excessive microbiological activity in water systems and to monitor the population over time and during remediation attempts.

Natural Organic Matter

Natural organic matter (NOM) reduces the corrosion rates of both galvanized steel and cast-iron pipe (Sontheimer et al. 1981, Larson 1966). NOM can also produce a more protective scale in iron pipes by reducing ferric colloids to soluble ferrous iron (Campbell and Turner 1983). One reason for this is that the presence of humic and fulvic acids promotes the formation of protective ferrous iron scales (Benjamin et al. 1996).

Schock et al. (1996) reported varied impacts of NOM on lead and copper solubility. In some instances, NOM was observed to form soluble organic complexes with lead, resulting in an increase in dissolved lead concentrations. Burlingame et al. (2006) noted that increasing NOM increased the occurrence of soluble lead. Similarly, Korshin et al. (2005) demonstrated that the presence of NOM can prevent or hinder the formation of lead carbonate species, resulting in increased lead concentrations. In other cases, NOM adsorbed or otherwise adhered to the inside of lead pipes, decreasing lead solubility.

Similar results have been reported for copper. Ferguson et al. (1996) reported that copper can form complexes with organic material, resulting in lower dissolved copper concentrations. Edwards et al. (1994a) noted that copper corrosion rates were substantially lower in the presence of 2 mg/L NOM than in its absence. Other researchers have discovered that NOM increases copper corrosion, particularly following one night of stagnation (Broo et al. 1999). In such a case, it is possible that MIC was occurring; however, this study did not investigate MIC as a possible cause of increased copper release.

Other Metals

Total metals and dissolved metals. When conducting metals analyses, it is important to distinguish between total metals and dissolved metals, for example, total lead and dissolved lead. The reasons associated with the occurrence of dissolved metals, as opposed to particulate metals, may be vastly different. Similarly, it is possible to see changes in the fraction of dissolved and particulate metals but see no change in total metals concentration. Cantor (2006) noted the importance of distinguishing between dissolved and particulate metals when trying to determine the potential source of corrosion-related problems. For example, a high fraction of particulate lead in a sample may indicate physically influenced metal release, such as scouring or water hammer, whereas a high fraction of dissolved lead tends to suggest chemically influenced metal release such as would result from stagnation in a lead service line (LSL). Boyd et al. (2004) found that colloidal lead concentrations represented 38 to 54 percent of total lead, and dissolved lead concentrations were 41 to 60 percent of total lead concentrations after stabilization in simulated partial LSL replacement studies. Deshommes et al. (2010) investigated sample collection methods and particulate lead presence in tap samples collected from homes with LSLs. The investigators found that there were two sources of particulate lead: (1) corrosion of lead-bearing elements in premise plumbing, which occurred at low to moderate concentrations with sporadic, higher levels; and (2) the adsorption of dissolved lead onto iron deposits present in the LSLs and premise plumbing, which occurred at low to moderate concentrations. Field surveys have indicated transient elevated lead levels immediately following partial service line replacement, which were attributed to physical dislodgement of particulate lead from the remaining lead service as a result of construction (Sandvig et al. 2008, McFadden et al. 2011).

Triantafyllidou et al. (2007) investigated whether particulate lead present in drinking water could be bioavailable to consumers while being undetected by the analysis method used for LCR compliance (USEPA 1994). The investigators found that, in most cases, the routine analysis procedure dissolved lead present in water samples and all samples had results above 15 μg/L regardless of method, demonstrating that the standard protocol successfully indicated problematic water quality. However, the investigators were able to point to unusual cases where particulate lead present in water appears to have contributed to lead exposure (Washington, D.C.; Greensboro, N.C.; and Durham, N.C.). According to the investigators (Triantafyllidou et al. 2007), the analysis method can underestimate lead levels if particulate lead is present due to

- particulate matter sticking to plastic sampling containers,
- using an inappropriate digestion method,
- inadequate dissolution because of particulate size and acidification time, and
- lead(IV) particles that dissolve more readily in simulated gastric fluid than in acidification during standard protocol.

Edwards and Dudi (2004) found that using a more aggressive digestion method resulted in lead levels that measured 500% higher than those samples analyzed using the USEPA standard protocol.

Madison Water Utility in Wisconsin exceeded the AL for lead in 1992 with a 90th percentile value of 16 μg/L. After it was determined that the source of the lead was LSLs and various treatment and other options were evaluated, the decision was made to implement an aggressive LSL replacement program. To evaluate the effectiveness of the program, first-draw samples were collected from 60 homes at which LSL replacement had been completed. Surprisingly, the 90th percentile value of these samples for total lead was 22 μg/L, or an increase of 6 μg/L over the pre-LSL replacement value. However, the 90th percentile value of the samples for dissolved lead was just 5 μg/L, leading Madison to conclude that particulate lead was the source of elevated lead levels (Cantor 2006). More recent visual, mineralogical, and chemical characterization of excavated LSL corrosion scales from Madison revealed that accumulated manganese (and iron) served as high-capacity sinks for lead. Manganese that accumulated from source well water onto pipe scales (up to 10 percent by weight of scale composition) captured and transported lead to consumer taps when the scale was disturbed (Schock et al. 2014b). After determining the problem to be particulate lead that was released both before and after LSL replacement, Madison warned consumers to continue allowing some water to run to waste before drinking for several years after LSL replacement. This example illustrates the importance in differentiating between total metals and dissolved metals concentrations, as well as monitoring for co-occurring substrate metals such as iron and manganese. It also shows that elevated lead levels can persist long after lead plumbing components have been removed.

Similar to the Madison experience, McFadden et al. (2011) observed order-of-magnitude short-term increases in particulate lead in residential samples taken after full replacement of LSLs. These increases were most pronounced in homes with galvanized iron plumbing, whereby the scales had served as a sink during historical exposure to the upstream service and acted as a source of lead after full service line replacement. In most cases, the short-term increase declined to pre-replacement levels or lower within a month. Continuing particulate lead release from remaining galvanized plumbing varied substantially from home to home, while dissolved lead release from the same galvanized pipes was negligible in most cases. The distinction between particulate and dissolved metals served as an important diagnostic tool for determining the causes and mechanisms of metals release in this study.

Aluminum. Aluminum may be a widespread component of distribution films and may act as a barrier to corrosion and metals release. Lauer and Lohman (1994) found that aluminum films significantly reduce lead leaching. Conversely, Scardina (2006) noted significant increases in the occurrence of copper pitting in plumbing systems that coincided with increased dissolved aluminum concentrations in the distribution system. Aluminum has also been found in conglomerated precipitates that appear to form around copper pinhole leaks (Rushing and Edwards 2002). However, a direct correlation between aluminum concentrations and the frequency of pinhole leaks has not been determined.

This parameter is also important in those systems that use aluminum-based coagulants, such as alum. An improper coagulation process may result in elevated aluminum levels entering the distribution system. Elevated aluminum levels become important when adding a phosphate-based inhibitor for corrosion treatment. Aluminum will readily bind to orthophosphate to form an aluminum phosphate precipitate ($AlPO_4$). Too much aluminum (>0.1 mg/L) may impede the ability of phosphate to react with lead and copper. In addition, aluminum phosphate can cause a precipitate that may increase "cloudy water" complaints at the customer tap (AWWA 2005). This precipitate is very small and has a low settleability rate, which can allow it to settle in low-flow areas at the ends of distribution systems.

Iron and manganese. In addition to the potential of iron and manganese to contribute to colored-water episodes, iron and manganese pose other concerns related to corrosion. Iron is an important parameter for several reasons. Some bacteria use iron as a food source, which can result in increased microbial activity and increased potential for MIC to occur (McNeill and Edwards 2001). In addition, both iron and manganese precipitates on pipe surfaces may form a physical barrier between the microorganisms and the drinking water disinfectant, allowing microbiological activity to intensify and thus cause an increased potential for localized corrosion (McNeill and Edwards 2001).

Recent studies have also shown that under certain conditions, manganese may be a significant component of lead pipe scale (Schock et al. 2008, 2014b). In one utility described previously, it was observed that manganese-rich scales had adsorbed lead and other metals. In this system, even after LSL replacement, particulates from manganese-rich scales were released, resulting in increased lead and other metals concentrations observed at consumers' taps.

Iron- and manganese-rich scales can accumulate lead within distribution system piping as well as customer plumbing that is not serviced by LSLs (Friedman et al. 2010, 2016; Scanlan 2003; Reiber and Dostal 2000). As described by Friedman et al. (2016), under certain conditions, the adsorptive capacity between a particular trace inorganic contaminant (TIC) and a solid (or common matrix element) can be such that the TIC's presence is considered to be "controlled" by the weight percent of the common matrix element. Figure 3-12 presents experimental results for the relationship between common matrix elements and TICs collected during main cleaning trials at two locations (AST and UPA) from a utility in Utah. The figure summarizes the R^2 value for specific TICs presence in relation to solid substrates (percent weight of the element). R^2 is used in Figure 3-12 to point out the correlation between matrix elements and contaminants, indicating how well a line of regression fit the dataset, with $R^2 = 1$ being a perfect fit. For each TIC, the substrate element that provides the strongest (positively sloped) correlation is shaded. These results suggest that manganese (Mn) solids exert a controlling effect on lead (Pb) and thallium (Tl), and to a lesser extent zinc (Zn), within the range of Mn occurrence experienced in these samples. Copper (Cu) control differed between the sites, which were supplied by different source waters (Friedman et al. 2016).

AST Site

Trace Contaminants	Common Matrix Elements: Fe	Mn	Ca	TUE
As	0.91	0.23	0.15	0.83*
Pb	0.30	0.90	0.00*	0.24*
Tl	0.01	0.86	0.09*	0.24*
Cu	0.83	0.46	0.13	0.77*
Zn	0.51	0.75	0.07	0.50*

* indicates negative slope

UPA Site

Trace Contaminants	Common Matrix Elements: Fe	Mn	Ca	TUE
As	0.28	0.05	0.01	0.00
Pb	0.13	0.72	0.03*	0.14*
Tl	0.03	0.84	0.07*	0.23*
Cu	0.29	0.90	0.01*	0.32*
Zn	0.04	0.59	0.20*	0.26*

* indicates negative slope

TUE = Total Unidentified Elements (e.g., biofilms and organics)

Source: *Friedman et al. 2016*

Figure 3-12 Linear regression R^2 values for paired occurrence data from ice pig trials at two locations (AST and UPA) from a utility in Utah

It is speculated that a reduction in ORP values may cause the manganese–lead scale to weaken or become soluble, increasing particulate lead concentrations (Schock et al. 2006, Maynard and Mast 2006).

As demonstrated by McFadden et al. (2011), galvanized iron plumbing may serve as a persistent source of lead. In some instances, lead release from galvanized plumbing was high enough to exceed the LCR AL even after other lead sources, such as LSLs, had been completely removed. Consistent with research conducted by others (Sandvig et al. 2008), the authors concluded that the lead release observed from the galvanized iron pipe was attributable to historical capture of lead released from upstream service lines.

Regulated trace metals. Corrosion scales and inorganic deposits, particularly those comprised largely of iron, manganese, aluminum, and/or phosphate, serve as substrates onto which various regulated trace metal and radiological contaminants (e.g., arsenic, lead, chromium, uranium) can concentrate via sorption/coprecipitation processes (Hill et al. 2010). This leads to the accumulation of metal contaminants within the distribution system and in premise piping/plumbing systems. Thus, the concern with corrosion is not limited to the metal that makes up the pipe but rather encompasses a wide array of inorganic contaminants that can effectively bind to the corrosion scale reservoir. Friedman et al. (2010) investigated the accumulation of 10 trace metals, uranium, and radium in 46 distribution system deposit samples collected from 20 utilities located throughout the United States. Samples represented a diverse array of system conditions and pipe types, including cast iron, cement lined, and plastic. Despite generally being non-detected in the water supplies, all 12 contaminants were detected in each of the deposit samples, indicating the universal nature of this phenomenon. Occurrence levels (in micrograms per gram, or μg/g) were contaminant specific and observed to span several orders of magnitude, depending on background water quality and deposit composition (Figure 3-13). Thus, risks can vary widely between systems and depend on system- and site-specific conditions. The most concentrated metals were lead (7,200 μg/g), barium (2,400 μg/g), and arsenic (900 μg/g). Lead and barium were generally correlated with manganese, indicating its role as a sink for trace metal cations.

In another study focused on accumulation of arsenic, Lytle et al. (2004) analyzed 67 deposit samples associated with varying pipe materials collected from 15 utilities in the Midwest. The researchers observed a mean arsenic occurrence level of 2,300 μg/g in hydrant-flushed samples and 900 μg/g in pipe-scale samples, with sorption/coprecipitation to ferric hydroxide solids/iron scale identified as the primary accumulation mechanism.

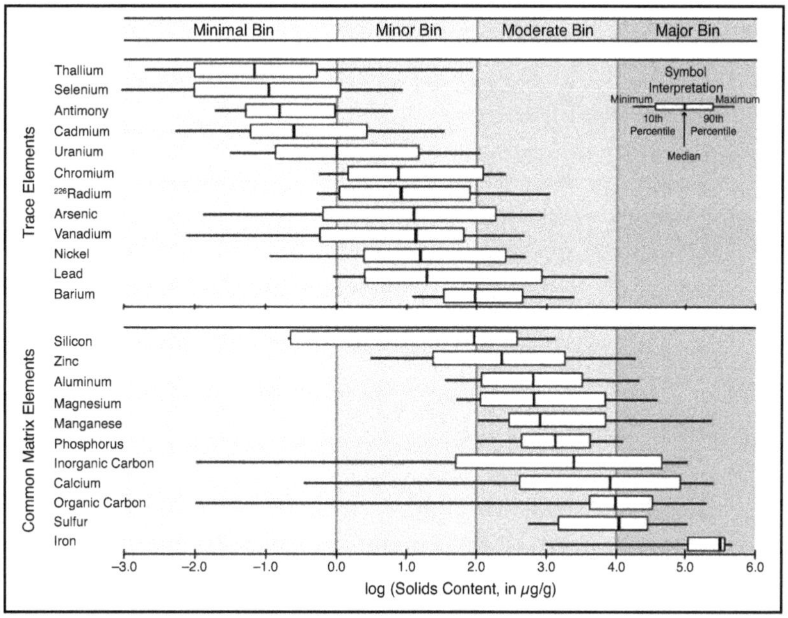

Source: *Friedman et al. 2010*

Figure 3-13 Occurrence distribution of trace and common matrix elements in distribution system deposits

Previously accumulated metals can be intermittently released back into the bulk water at elevated concentrations, presenting an acute or sub-chronic public health risk (Schock 2005, Hill et al. 2010, Friedman et al. 2016, Schock et al. 2014b). Releases can occur as a result of routine system operating conditions, including hydraulic mobilization of coprecipitates because of changes in flow velocity or direction, as well as chemical reequilibration processes (e.g., desorption, dissolution) due to changes in supply, blend, and/or water chemistry. With trace contaminants such as arsenic being present as high as parts-per-hundred levels in certain cases, fractional deposit releases can result in significant bulk water concentration increases at customer taps with no obvious visual discoloration or indication of upset (Hill et al. 2010). The current regulatory monitoring and compliance framework for inorganic contaminants requires monitoring only at the entry point to the distribution system, with the exception of lead and copper that are monitored at the tap, so many utilities do not have data on contaminant levels beyond the distribution system's entry point. However, several recent investigative monitoring efforts have demonstrated at-the-tap metals excursions of existing maximum contaminant levels, action levels, and/or health advisory levels for acute exposure scenarios specifically attributed to accumulation and release. Notable examples from published case studies include the following at-the-tap results: copper at 300 mg/L, arsenic at 5 mg/L, lead at 1,400 μg/L, and radium at 12.5 pCi/L (Hill et al. 2010).

Turbidity and Color

Turbidity is the measurement of the amount of light diffraction in a sample caused by particulates and sediments; therefore, the higher the turbidity, the higher the level of particulates in the water. These particulates may cause water discoloration or eventually settle on pipe walls, providing increased opportunity for microbial regrowth in the distribution system and accumulation of other constituents (Post et al. 2011). Apparent color compares the color of water to that of deionized water.

Turbidity and color are relevant to corrosion for several reasons. Both will increase during red-water or other aesthetic events. In addition, these parameters, combined with total metals and dissolved metals analyses, may help to better characterize a corrosion-related event. For example, an increase in the number of red-water complaints accompanied by an increase in turbidity and particulate iron but with no significant change in total iron concentrations may indicate that something is occurring in the distribution system to oxidize ferrous iron already present. Furthermore, increases in turbidity caused by iron particles may also result in increases in particulate lead concentrations because of adsorption of lead onto the iron scales (Kirchner 2009). This data, when combined with chlorine residual, ORP, and total iron and phosphate (total phosphate, orthophosphate, and/or polyphosphate), may help to understand why the red-water event occurred.

Other Considerations

In addition to water quality, many other factors not only influence corrosion and metals release but can also indicate corrosion-related problems. Hydraulic factors, customer complaints, and non-revenue water are useful considerations in identifying the location and cause of water quality problems related to corrosion.

Hydraulic factors. Several researchers have observed that corrosion rates of lead, copper, and iron are likely to be highest in areas of the distribution system with the longest hydraulic residence time, or water age (Edwards et al. 1994a, Imran et al. 2005). Several reasons are possible, the most significant being that disinfectant residuals tend to be lower in areas of the distribution system with higher water age. Theoretically, low-disinfectant residuals can reduce the ORP of the water and can cause shifts in metal species in existing pipe scales, resulting in an increase in dissolved metals concentrations. Further, low or nonexistent disinfectant residuals increase the potential for MIC. It has been observed that low-flow or stagnant areas of the distribution system or home plumbing systems are more prone to MIC (Cantor et al. 2003, Bremer et al. 2001).

It has also been observed that stable iron corrosion deposits may be disrupted by changes in flow (Smith et al. 1998). Pisigan and Singley (1987) found that mild steel and copper corrosion increased significantly as flow rates increased. It was observed that with time, passivating layers on the pipes could form; however, further increases in the flow rates or turbulent flow conditions (as are typically found in distribution systems) may disrupt the protective surface. Corrosion rates of copper piping are not anticipated to be excessive if the flow rates are less than 4 to 5 ft/s (Obrecht and Quill 1960, 1961).

Water quality customer complaints. Changes in aesthetic parameters, identified via monitoring or increases in customer complaints, may indicate that corrosion control treatment effectiveness has been compromised or that a metal release event has occurred. For example, customers may notice a change in taste, odor, or color in old or stagnant water. In Park City, Utah, the Park City Municipal Corporation (PCMC) experienced two "release" events in 2007 and 2010, which caused numerous discolored-water customer complaints and resulted in temporary excursions for several metals in two service areas of the distribution system. PCMC has embarked on aggressive monitoring and main cleaning programs as well as increased controlled blending of sources and chemical stability of finished water to address these issues (De Haan and Friedman 2015, Friedman et al. 2016). It is important to note, however, that metals released in the soluble form may not result in visible discoloration yet still could cause exceedances (Hill et al. 2010).

Non-revenue water. Non-revenue water, or water losses, may also be an indicator of widespread corrosion control failure. Significant increases in non-revenue water may result from failure of distribution system piping and home plumbing systems. In Brown Deer, Wis., an increase in non-revenue water from 3.4 percent to 9.4 percent over a four-year period coincided with a substantial increase in the occurrence of copper pipe failures in home plumbing systems (Cantor et al. 2003). Subsequent evaluation of the system, after remedying the copper pipe corrosion problem, showed non-revenue water had dropped to 2 percent (Cantor et al. 2006).

PART 2: DEVELOPING AN INVESTIGATIVE WATER QUALITY MONITORING PROGRAM

This section focuses on corrosion-related considerations for water quality monitoring with the objective of understanding baseline (current) conditions and factors contributing to or controlling corrosion processes. Chapter 6 provides a more detailed discussion of ongoing monitoring that can be used to demonstrate optimization of corrosion control as a result of treatment implementation, modification, or modified operational strategies. As discussed previously, internal corrosion and deposition processes can present additional distribution system water quality challenges pertaining to trace metals accumulation and release. Additional monitoring concepts relevant to characterization of this phenomenon are also presented in this section.

Importance of Establishing Baseline Water Quality

It is necessary to establish baseline water quality not only to identify the conditions under which corrosion control is most effective but also to determine the cause of corrosion-related episodes, such as red-water events and increased tap lead concentrations. Baseline water quality data should be established for the source water, treatment plant, entry point, and distribution system. Typically, there is too much variability among individual homes because of home age, plumbing age, configuration, material types, and water use habits to establish a meaningful baseline for household plumbing conditions.

Many systems may have historical data that can help provide initial baseline data when a monitoring program is initiated. Many times, however, the historical data contains only a few water quality parameters collected for treatment optimization or regulatory requirements. In most cases, baseline monitoring will need to be extended to a host of other water quality parameters collected at locations that are not traditionally common for routine sampling. Additional tests on distribution system pipes, such as X-ray diffraction (XRD) and scanning electron microscopy (SEM) analyses, can provide additional information related to the physical makeup of the pipe scale surface but are not required as a part of a routine monitoring program. These methods are discussed in greater detail in chapter 5.

Monitoring Parameters for Corrosion Control

Routine water quality monitoring is useful for ensuring the effectiveness of, as well as identifying the potential cause of upsets in, corrosion control treatment. Table 3-3 provides a summary of recommended water quality monitoring parameters. Level 1, or baseline, parameters should be monitored at routine intervals at the entry point and throughout the distribution system, for example, daily and monthly, respectively. Level 2 parameters are supplemental and may be measured in conjunction with Level 1 parameters if time and budgets allow, or Level 2 may be measured for the purposes of verifying that optimal conditions for corrosion control effectiveness have not been compromised. It may be helpful to first establish baseline concentrations of Level 2 parameters before deciding only to monitor for these parameters on an intermittent basis. Level 3 parameters may be used to pinpoint the probable cause of corrosion problems after the corrosion occurs and do not need to be monitored on a regular basis.

The collection of any supplemental tap water samples should be carefully reviewed in advance with the authority having jurisdiction to appropriately distinguish between

Table 3-3 Suggested water quality monitoring parameters

Level 1 (Baseline)	Level 2 (Supplemental)	Level 3 (Diagnostic)
pH	Temperature	Bacterial speciation
Total alkalinity	Hardness (total and calcium)	Sulfide
Dissolved inorganic carbon (measured)	Conductivity	Nitrite–nitrate*
Phosphate (total and ortho-, if used)	Dissolved oxygen	
Disinfectant residual	Oxidation–reduction potential	
Ammonia (free and total)*	Lead (total and dissolved)§	
Chloride	Copper (total and dissolved)§	
Sulfate	Iron (total and dissolved)	
Aluminum†	Manganese (total and dissolved)	
Iron (total and dissolved)‡	Heterotrophic plate count/adenosine triphosphate	
	Color (apparent)	
	Total organic carbon, total nitrogen, total phosphorus	
	Turbidity	

*Systems practicing chloramination only.

†Systems practicing alum or powdered activated carbon coagulation or those systems with high source water aluminum concentrations.

‡Systems using ferric salts for coagulation, systems with a high percentage of iron mains, and those with high source water iron concentrations.

§Lead and copper samples may be required to be reported by the state or federal primacy agency in the United States. The implications of additional lead and copper sampling should be discussed prior to sample collection and analysis.

(1) samples collected for compliance and comparison of regulated levels (e.g., LCR) and (2) those efforts designed for the purposes of maintaining effective corrosion control treatment and understanding influencing mechanisms. Utilities should also review their public communication policies and decide how results of supplemental sampling will be communicated to those customers and others that participate in the study.

Frequency of Monitoring

For optimal corrosion control treatment, USEPA requires utilities to monitor only a few parameters, for example, pH, alkalinity, hardness, and phosphate (if an inhibitor is added) at the entry point once every 2 weeks and twice every 6 months at locations in the distribution system. Reduced monitoring decreases distribution system sampling considerably (*Federal Register* 1991).

Compliance monitoring alone is insufficient for adequately characterizing system conditions and monitoring corrosion control effectiveness. A monitoring frequency of every six months or annually will not be able to capture water quality conditions that could change because of seasonal changes in source water quality, temperature, and consumption. USEPA's most recent technical recommendations for evaluating optimized corrosion control (USEPA 2016) recommend follow-up water quality parameters at a monthly frequency with the objective of observing seasonal water quality conditions in the distribution system. It is recommended that utilities develop and implement an operational monitoring program to assist in optimizing corrosion control treatment. Arrangements of the operational monitoring program must be worked out with the appropriate regulatory agency to understand what portion, if any, of operational data may be reportable.

Entry-point readings should be taken more often than required for lead and copper compliance monitoring, preferably on a daily basis. For systems with multiple entry points (e.g., groundwater systems with multiple well fields), a program under which each entry point is monitored routinely is still recommended; however, daily monitoring may not be practical or economical. In such cases, weekly monitoring may be more practical for several reasons. The importance of entry-point monitoring is to ensure that water quality entering the distribution system is consistent and stable. Small changes in source water pH, alkalinity, and DIC may impact corrosion control treatment (i.e., water quality parameters) leaving the plant. Increased distribution monitoring frequencies can determine areas where water may stagnate or identify areas influenced by MIC. On-line analyzers are available for multiple parameters, including chlorine, pH, turbidity, and others. Chapter 6 provides more detail on different monitoring approaches and strategies.

After baseline levels have been established, monitoring of Level 2 and Level 3 parameters may be substantially reduced. However, monitoring frequencies should increase whenever there is a change in source water, treatment, or distribution practices that impacts water quality or physical factors that influence corrosion. For example, increased monitoring of tap lead concentrations (total and dissolved) is recommended following a change in pH or phosphate treatment. Similarly, a utility might conduct ORP, lead, and nitrification monitoring before and after a change from chlorine to chloramine.

Monitoring Locations

Baseline water quality parameters should be monitored at the source, entry point, and select locations in the distribution system. In addition to routine baseline monitoring, a period of more frequent and targeted monitoring in the distribution system may be the only practical method of gathering data on in situ materials and conditions. Such a targeted monitoring program would include selected Level 2 and Level 3 parameters.

In the United States, required LCR monitoring targets single-family homes that contain copper pipes with lead solder installed after 1982 or lead pipes and/or are served by an LSL (*Federal Register* 1991). When insufficient single-family homes are available to meet this requirement, other buildings, including multifamily dwellings, may be sampled. It has been demonstrated, however, that lead and copper problems may occur in plumbing of both new and old homes. Kimbrough (2007) found that brass corrosion can be a major contributor to tap lead and copper concentrations, even in homes with plastic plumbing. Cantor (2000) demonstrated that new copper plumbing may, in fact, be more susceptible to corrosion than older copper plumbing. Homes with new or replaced meters and/or LSLs may also be more susceptible to lead and copper problems. Meter and service line replacement may result in particulate lead and copper originating from the service lines (Sandvig et al. 2008) and mobilized by construction disturbances (Del Toral et al. 2013).

Partial replacement of LSLs may result in galvanic corrosion in which the lead pipe itself becomes the sacrificial anode. Therefore, it is critical to establish representative metal release rates before, during, and after replacement.

For the reasons previously described, utilities may want to consider a supplemental operational tap water testing program. Such a program should carefully consider the following:

- Sampling locations may need to be expanded to include conditions that represent homes with new copper plumbing, brass faucets, new or replaced meters, and/or replaced or partially replaced LSLs. Ideally, these new locations would be homes or other building locations under the utility's direct control.
- Utilization of Total Coliform Rule monitoring locations to assess the stability of distributed water quality and possible impacts on corrosion may also be useful.

Utilities should review the Level 1, Level 2, and Level 3 parameters previously described to determine the most appropriate parameters for their system.

- Sample collection from some homes should be done in multiple consecutive 1-L samples: for example, first liter, second liter, third liter, initial temperature change, 5 minutes, and so on. Smaller consecutive aliquots may be appropriate depending on the volume of plumbing in the home. Collection of consecutive samples collected from an internal tap to the service line (or other end point) is known as *profiling*. LCR compliance is based on the first-draw sample (i.e., first liter of water flushed from the tap) based on the assumption that the first-draw sample represents the worst-case sample. However, when LSLs or brass fittings are present within the plumbing system, samples with higher lead levels may be observed in subsequent pipe volumes. Because of the time and expense involved, utilities may want to consider profiling only in certain situations (e.g., following LSL replacement or in the event a change in tap water lead or copper concentrations is observed). Profiling is discussed later in this chapter and in more detail in chapter 6.

Monitoring Concepts for Metals Accumulation and Release

Trace metals accumulation and release within utility distribution and customer premise plumbing systems involves several complex, path-dependent, and dynamic processes that are impacted by a host of system and site-specific variables (USEPA 2006). These include past and current metals loading, water quality, and chemistry; pipe type, age, condition, and maintenance history; and hydraulic conditions. Given these numerous variables, distribution system monitoring remains the most useful and direct method to characterize metals accumulation and release at the local level. Because of the lack of current federal regulation or monitoring requirements surrounding this issue, such monitoring endeavors are voluntary and investigative and should be communicated as such to the regulatory agency and any local laboratories.

Utilities are encouraged to develop and implement an investigative monitoring program that builds upon a utility self-assessment. Friedman et al. (2010) provide recommended content for such an assessment. It involves a series of conditions and observations that suggest risk. By undergoing this process, utilities can focus their monitoring efforts and more reliably determine what, where, and when to monitor. The importance of using an initial assessment to tailor the monitoring effort is underscored by the complex and expensive nature of several of the analytical techniques involved in characterizing trace metals.

Monitoring programs should involve two basic components: (1) collection and analysis of native system deposits, and (2) implementation of a carefully designed bulk water monitoring program to assess release and potential exposure conditions (Friedman et al. 2016, Hill et al. 2010). The collection and analysis of native solids is critical to understand which contaminants are present, where they occur, and at what concentrations. Since accumulation is a precursor for release, identification of spatial trends in contaminant occurrence and inventory provides a basis for focusing bulk water monitoring efforts to identify mobilization behavior. Native deposits can be obtained from pipe specimen samples, including dedicated extractions and samples of opportunity (e.g., from main breaks, pipe taps) as well as during flushing trials. Debris from tank cleaning can also be analyzed, but care should be taken because materials accumulated within storage facilities may not be representative of reactions and accumulation on pipe surfaces. With regard to monitoring locations, an initial assessment should provide broad spatial coverage, with added emphasis in areas that present accumulation risk factors (e.g., areas routinely served by sources with measureable concentrations of trace metals, iron, manganese, aluminum, and/or phosphate; areas with unlined iron pipe; low-flow areas; areas of customer water quality complaints or colored water).

A carefully devised bulk water monitoring program can provide insight into trace metals transport behavior in the system. Concurrent with assessment of the distribution system, it is important to establish a detailed baseline understanding of entry-point water quality and chemistry conditions for each active source and develop an understanding of system blending regimes. As stated previously, speciation of particulate and dissolved metals is important to assess release, understand the mechanism, and identify causative factors. With regard to locations and timing, following a period of baseline monitoring, emphasis should be placed on areas of the system that present an increased risk of release (e.g., areas of significant accumulation, before and after source use and/or treatment changes; areas that receive supply from multiple, dissimilar sources; in response to customer complaints or discolored water; during and after flushing; in response to main breaks; near pump station operations; over a 24-hour diurnal cycle, including during peak demands; and in areas of high water age). With regard to parameters, Friedman et al. (2016) established a general framework intended to help utilities identify release events, the mechanism of observed releases, and potential causes. Having an understanding of the mechanism and cause of the release, in addition to the magnitude and location, can allow utilities to implement specific strategies to eliminate or otherwise mitigate release risks. The framework developed by Friedman et al. (2016) allocates parameters into four groups with specific purposes, as follows:

1. **Source tracking parameters** (for systems with multiple sources that blend within the distribution system). This includes conservative parameters used to trace which source(s) supplied each location and the approximate blend fraction (e.g., electrical conductivity, chloride-to-sulfate ratio, and TDS).
2. **Trace metals.** This can include the entire universe of regulated inorganic contaminants or a more focused subset based on historical and current water quality conditions. Particulate and dissolved fractions should be determined using the lowest detection limits possible.
3. **Substrate elements.** Certain elements form relatively insoluble compounds that serve as highly effective substrates for adsorption and coprecipitation of trace metals (e.g., iron, manganese, and HPC-R2A [biofilm], or turbidity and color as indicator parameters). By obtaining information on these parameters, it may be possible to develop an understanding of local accumulation and release mechanisms and identify correlations that can provide rapid identification of potential hydraulic and physical releases.
4. **Water chemistry.** Water chemistry parameters impact precipitation-solubility and sorption–desorption equilibria (e.g., pH, temperature, alkalinity, disinfectant residual, DO, ORP, hardness, TOC).

CHARACTERIZING CORROSION CONTROL AND METALS RELEASE

This section provides guidance regarding how to use water quality and other data to determine the most likely cause of internal corrosion and metals release in a distribution system. As discussed in chapter 2, many mechanisms may result in increased metals concentrations in drinking water distribution systems and home plumbing systems—various types of uniform corrosion, nonuniform corrosion, and non-corrosion mechanisms. For convenience in diagnosing a corrosion issue, these mechanisms are reclassified here as chemically influenced, microbiologically influenced, and physically influenced.

Being able to quickly and accurately determine the cause of a corrosion-related incident—for example, red water or increases in tap lead concentrations—first requires that baseline water quality be established. This baseline may then be compared to subsequent water quality monitoring results to help pinpoint the cause of the event. For example,

- Is there a correlation between distribution chlorine residual data and the occurrence of home plumbing failures? Such a correlation could be an indicator that MIC is occurring.
- What was the impact of a change in coagulant on the finished water chloride-to-sulfate ratio?
- What impact did treatment changes have on finished water pH and alkalinity?
- What impact did a new source of supply have on the stability of blended water?
- Is there a need to modify the current corrosion control strategy?

Not only can water quality data help to determine the cause of a corrosion-related event, but they can also be used to eliminate other possibilities. That is, if there were no significant changes in key corrosion-related water quality parameters, there may have been physical factors that led to the event. For example, if there were no changes in water quality in an area of the distribution system that recently experienced an unprecedented number of red-water complaints, some of the other potential factors would need to be ascertained. Were there system maintenance activities in the area that might have contributed to the event? Were there changes in flow (velocity or direction) or system pressure that might have caused shearing of existing pipe scale? Is there a storage tank nearby? Is that tank well mixed? Is it possible that water discharged from that tank was of poor quality and contributed to the event? Has the tank been cleaned recently? Is it possible that sediment or other debris may have come from the tank? Does the location of the corrosion issue correspond to a part of the distribution system that undergoes fluctuations in water quality as a result of uncontrolled blending (i.e., the pipes are exposed to alternating sources and blends thereof)?

Recognizing locational patterns can provide important information to contextualize water quality data. Physical, chemical, or microbial influences on corrosion may be prevalent in a particular area due to unique system configuration or operation. Hydraulic system models are becoming commonplace, but typically there is not an associated water quality component of such models. More accessible tools are available that utilities can employ in their own analyses of corrosion issues. Geocoding of data using a simple geographic information system (GIS) platform, especially customer complaints or LCR monitoring data, provides a simple means of identifying locational issues. Once a location is identified, it can facilitate more focused analysis of water quality data specific to that locale to diagnose the cause of the corrosion issue. For example, in a distribution system evaluation at the City of Rochester, N.Y., to determine the cause of a long-term increasing trend in LCR 90th percentile monitoring data, McFadden et al. (2011) used geo-referenced LCR data to identify an area of the system where residential lead samples were markedly higher than elsewhere, contributing substantially to the increasing 90th percentiles. Further analysis of system hydraulics indicated that this area corresponded geographically to the region where two source waters were most likely to blend. Water quality analysis of the two sources revealed divergent trends in pH and alkalinity (i.e., the carbonate system impacting lead solubility and scale morphology) between the two sources, increasing the magnitude of water quality fluctuations experienced in pipes in this area. These fluctuations in the carbonate system were likely diminishing the stability of passivating scales, leading to increased lead release. This study illustrates the importance of locational and hydraulic factors in understanding and addressing corrosion-related issues. Tools such as GIS for spatial analysis or spreadsheet models for analysis of corrosivity characteristics can be readily employed at minimal expense.

Table 3-4 provides a list of common corrosion-related water quality problems and the potential causes of those problems in the context of three general categories. The list is by no means exhaustive and can be further broken down into smaller categories as described in chapter 2. The list is intended to be used as a guide in conjunction with the water quality considerations discussed in this chapter when assessing water quality and other impacts related to corrosion.

Examination of Pipe Scales

In many cases, existing corrosion-related issues and/or metals behavior cannot be easily understood or explained through bulk water quality monitoring alone. In such cases, the collection and direct examination of pipe specimens, corrosion scales, or native accumulated deposits can often provide useful insight into the underlying mechanism and the appropriate mitigation or response strategy. Approaches to characterization can be sorted into four categories of analysis, which are discussed below: visual, elemental, mineralogical/morphological, and microbial. Schock and Lytle (2011) indicate that investigating pipe scales by dissolving them in acid and conducting an analysis to determine presence of potential metals or compounds can provide an overview of the type or cause of corrosion. An evaluation including samples from multiple locations within the distribution system can indicate pipe materials that are experiencing relatively more corrosion (Schock and Lytle 2011).

Figure 3-14 provides a general overview of processing and analytical techniques available for various types of solid materials associated with drinking water distribution systems. The applications value of these techniques are described in more detail below. Since the techniques provide different types of information on solid phases, it is common to use multiple techniques to obtain complementary insights.

Regarding sample processing, in some instances, pipe scales and deposits exhibit a nonuniform structure such that it may be insightful to harvest solids in distinct layers as feasible, based on differences in appearance (color and texture) and tenacity. Differences across the depth of a scale may provide useful information on conditions affecting corrosion (near the pipe surface) and solubility (near the water interface).

Visual Analysis

The opportunity to see the interior surface of a particular pipe can often provide powerful diagnostic information. Basic macroscopic pipe and/or scale properties such as color, texture, thickness, layering, pitting, and so forth, can be used to support or eliminate specific hypotheses, compare with known or reference conditions, or otherwise be used to guide subsequent evaluations.

For example, Schock et al. (2014b) were able to observe notable differences in the appearance and tenacity of in situ scale in harvested LSL specimens (see Figure 3-15) that provided insight into lead release mechanisms. One specimen with a protective platternite layer (as confirmed by XRD) was dark reddish brown, relatively thin, and firmly attached, whereas another sample from the same utility had a black outer scale that was thicker and easily flaked off to reveal a yellow layer underneath. Additional chemical characterization revealed that the outer layer of this second specimen contained high levels of manganese with sorbed lead, which, in addition to serving as a source of particulate lead release, also contributed to higher plumbosolvency by preventing the formation of an insoluble PbO_2 layer. Laboratories that specialize in materials analysis are often able to provide these types of services.

Table 3-4 Assessment of common corrosion-related water quality problems

	Indicators That the Symptom May Be . . .		
Symptom	Chemically Influenced	Microbiologically Influenced	Physically Influenced
Red water	• Shift in dissolved/ particulate iron speciation but no change in total iron concentration • Loss or reduction in disinfectant residual • Change in oxidation–reduction potential (ORP) • Shift in pH • Change in distributed water alkalinity • Treatment process upsets • Change in finished water chloride-to-sulfate ratio	• Loss or reduction in disinfectant residual • Increase in heterotrophic plate count (HPC)–R2A or adenosine triphosphate (ATP) • Localized shifts in pH but not system-wide • Increases in distribution nitrite–nitrate concentrations* • Higher or lower than normal free ammonia concentrations* • Higher or lower dissolved organic carbon (DOC) concentrations	• Changes in flow • Stagnation • Changes in system pressure • System maintenance • Release of sediments from distribution storage tanks • Release of lower-quality water from poorly mixed storage tanks • Distribution storage tank cleaning
Increased tap lead concentrations	• Recent change in distribution system disinfectant • Change in ORP • Drop in finished water pH • Shift in pH to outside optimal range for corrosion control chemical • Change in distributed water alkalinity • Change in finished water chloride-to-sulfate ratio • Upset of phosphate addition system resulting in lower phosphate concentrations • Poor buffer intensity	• Loss or reduction in disinfectant residual • Increase in HPC-R2A or ATP • Localized shifts in pH but not system-wide • Increases in distribution nitrite–nitrate concentrations* • Higher or lower-than-normal free ammonia concentrations* • Higher or lower DOC concentrations	• Increase in particulate lead concentrations but no corresponding increase in dissolved lead concentrations • Recent lead service line replacement • Major disturbances that shake service lines or plumbing • Hydraulic transients • Installation of new brass plumbing fixture • Long periods of stagnation
Increased tap copper concentrations	• Recent change in secondary disinfectant • Change in ORP • Drop in finished water pH • Shift in pH to outside optimal range for corrosion control chemical • Change in distributed water alkalinity • Change in finished water chloride-to-sulfate ratio • Upset of phosphate addition system resulting in lower phosphate concentrations • Introduction of a hard, high alkalinity (dissolved inorganic carbon) groundwater	• Loss or reduction in disinfectant residual • Increase in HPC-R2A or ATP • Localized shifts in pH but not system-wide • Increases in distribution nitrite–nitrate concentrations* • Higher or lower-than-normal free ammonia concentrations* • Higher or lower DOC concentrations • Presence of sulfide	• Increase in particulate copper concentrations, but no corresponding increase in dissolved copper concentrations • Long periods of stagnation

Table continues next page

Table 3-4 Assessment of common corrosion-related water quality problems (continued)

	Indicators That the Symptom May Be ...		
Symptom	Chemically Influenced	Microbiologically Influenced	Physically Influenced
Pinhole leaks in copper plumbing	• pH below 7.2 • High chloride relative to sulfate and bicarbonate • Treatment upsets resulting in increased residual aluminum concentration in finished water	• Loss or reduction in disinfectant residual • Increase in HPC-R2A • Localized shifts in pH but not system-wide • Increases in distribution nitrite–nitrate concentrations* • Higher or lower-than-normal free ammonia concentrations* • Presence of sulfide • Higher or lower DOC concentrations	• Stagnation • Sediment or debris in plumbing • High velocity (erosion corrosion) • Improper fluxing (leads to "tracks" on inside of pipe)
Blue water (note: can occur in newly installed or older plumbing and service lines)	• Drop in pH below 7.2 in soft waters • Rise in pH above 8.0 in soft waters • Drop in finished water alkalinity	• Loss or reduction in disinfectant residual • Increase in HPC-R2A • Localized shifts in pH but not system-wide • Increases in distribution nitrite–nitrate concentrations* • Higher or lower-than-normal free ammonia concentrations* • Higher or lower DOC concentrations	• Stagnation
Pink water	• pH shift • Treatment upsets resulting in increased MnO_4^- concentrations entering distribution system	• Pink water is not likely to be the result of microbiologically influenced corrosion (MIC)	• Changes in flow • Stagnation • Changes in system pressure • System maintenance • Release of sediments from distribution storage tanks • Release of lower-quality water from poorly mixed storage tanks • Distribution storage tank cleaning
Black or yellow water	• Drop in finished water ORP • Treatment upsets resulting in reduction of manganese dioxide (MnO_2)	• Black or yellow water is not likely to be the result of MIC	• Changes in flow • Stagnation • Changes in system pressure • System maintenance • Release of sediments from distribution storage tanks • Release of lower-quality water from poorly mixed storage tanks • Distribution storage tank cleaning

*Systems adding chloramine.

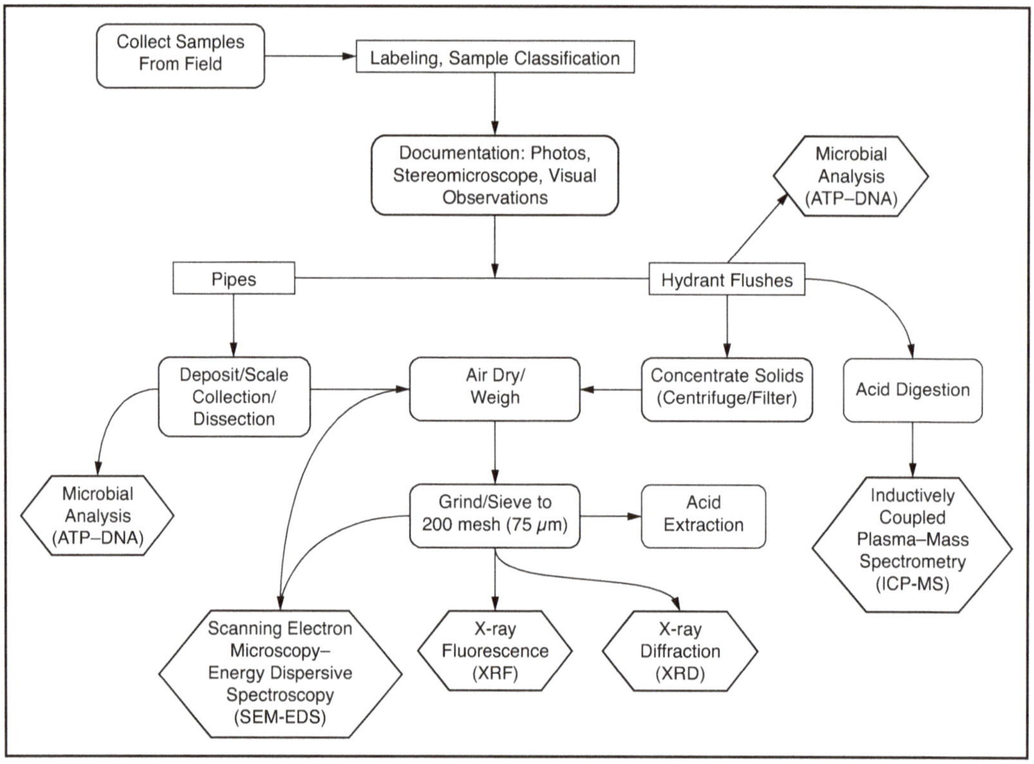

Adapted from Lytle et al. 2004

Figure 3-14 Overview of solid sample processing and analytical techniques

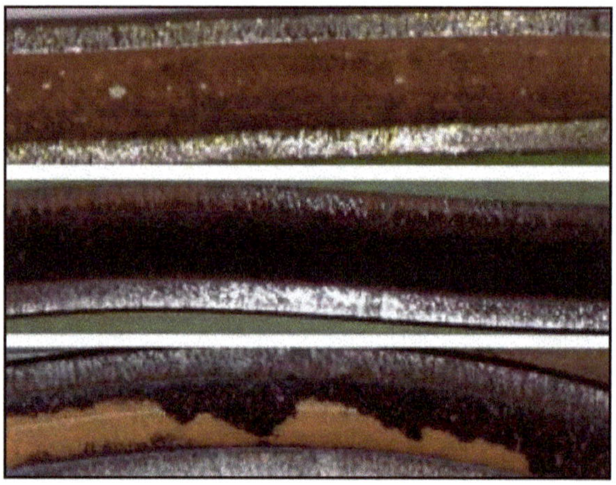

Source: *Schock et al. 2014b*

Figure 3-15 Longitudinal cross sections of lead pipe showing differences in corrosion scales

Elemental Analysis

The inorganic elemental composition of harvested, pulverized scale can be determined through a destructive process that involves application of a solids extraction process (e.g., USEPA Method 3050B; 1996) to solubilize the solid matrix followed by analysis of the aqueous extract using USEPA or standard methods. This produces data on the concentration of targeted inorganic elements in the matrix, expressed as micrograms per gram of solid (parts per million) or weight percent.

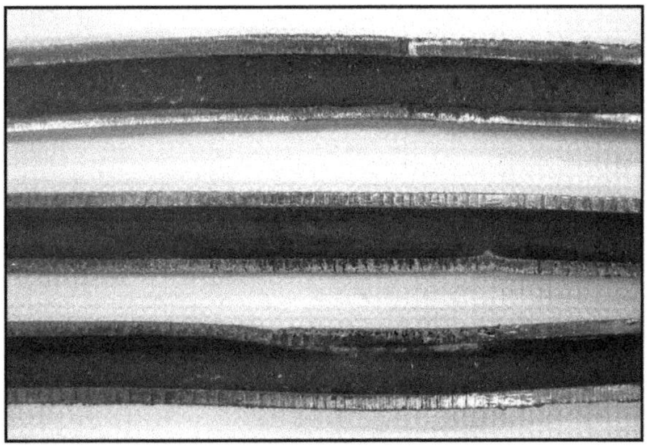

Courtesy of USEPA

Figure 3-16 Pipes with lead-dioxide scales cut horizontally prior to XRD

Elemental analysis is particularly useful to assess occurrence of trace metals accumulation in distribution system deposits, co-occurring substrates (e.g., iron, manganese), as well as the makeup of solid layers on metal service lines and plumbing components. By its nature, the process provides aggregate results (i.e., samples are crushed and homogenized in advance, although discrete layers can be processed separately), which can be compared with surface analysis and elemental mapping (scanning electron microscopy–energy dispersive spectroscopy, or SEM-EDS) to better understand contaminant distribution, layering, and potential mobility.

X-ray fluorescence (XRF) can also be used to conduct elemental analysis that requires a simple sample preparation in comparison to those requiring wet chemical analysis. The sample is ground into a fine powder, then formed into a pellet or fused into a glass bead. The accuracy of this method is comparable to techniques requiring dissolution for constituents down to the order of high-parts-per-million levels by weight. However, XRF requires a large sample relative to the quantity of material typically found in lead and copper piping (Schock and Lytle 2011).

Mineralogical/Morphological Analyses

X-ray diffraction (XRD) is a procedure used to identify the composition of specific mineralogical phases in a solid sample. In the process, X-rays are scattered by crystal atoms on the pipe wall (or in a powdered composite of the scraped scale), producing a diffraction pattern that yields information about the crystalline composition. The process cannot be used to identify amorphous solids, such as hydrous manganese oxides. To conduct XRD for corrosion pipe-surface information, typically 1-ft sections of service lines or internal copper plumbing pipes are removed from the system. Pipe specimens are cut longitudinally with a band saw having a fine metal-cutting blade (Schock and Giani 2005) (Figure 3-16). Scales are then carefully removed to conduct the analysis. Figure 3-17 illustrates XRD patterns for lead-oxide scales.

Scanning electron microscopy (SEM) can be used to visualize solids morphology by producing high-magnification micrographs (100–500×). Images can be compared to identify notable differences between multiple samples as well as to reference solids of a known composition to identify probable solid phases. Energy-dispersive spectroscopy can be used in conjunction with SEM to identify surface composition. In a process known as elemental mapping, discrete user-selected locations on a deposit (as viewed in SEM) can be analyzed to obtain surface elemental composition (Lytle et al. 2004, Schock and DeSantis 2014). This is particularly useful for solids with highly diverse morphological

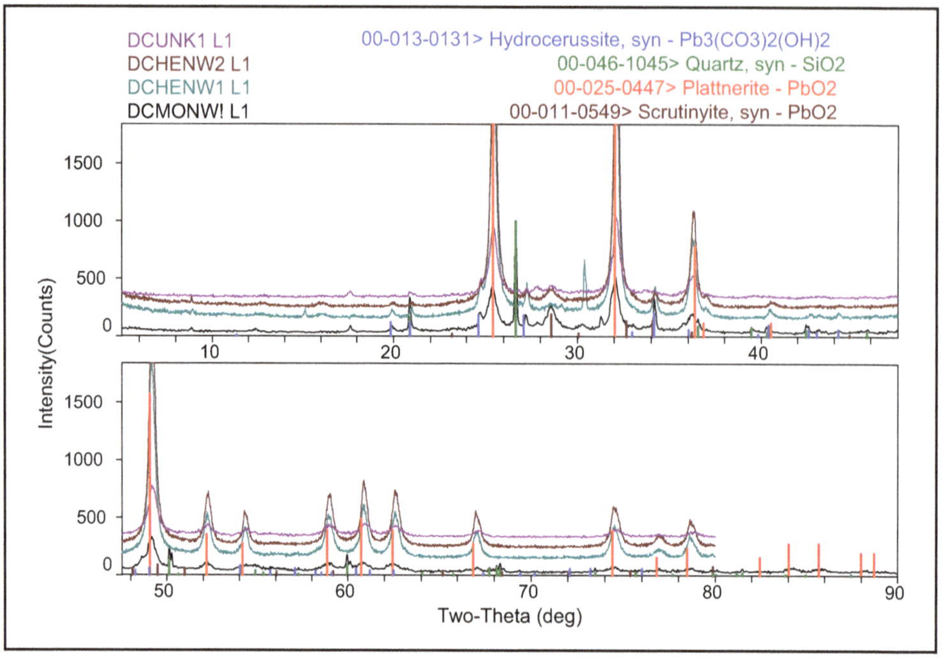

Courtesy of USEPA

Figure 3-17 XRD pattern for an LSL from Washington, D.C.

features and/or as a basis of comparison with elemental analysis of multilayered or thicker scales (>1 mm) where surface properties may differ substantially from the aggregate harvested deposit.

Conducting XRD and SEM analysis on pipe wall scales can provide supplemental information regarding existing scale formation (AwwaRF and TZW 1996, Lytle et al. 2004, Schock and DeSantis 2014). This data can provide a picture of historical effects from past treatment and water quality effects on pipe walls. The data can also provide insight into how a scale might react to a change in drinking water treatment or source water change. For example, in the District of Columbia, when lead compliance samples were exceeding USEPA action levels, XRD samples from LSLs removed from residents' homes revealed thick scales of lead dioxide (PbO_2). This finding helped experts understand the structure of the scale and its lead-leaching potential into the drinking water. Since the introduction of orthophosphate as treatment to the lead-oxide scale, XRD samples have displayed slow changes of the lead oxide to lead phosphate, providing information to the utility that the scale is becoming more stable and that the lead is transitioning into a form that is less soluble.

More recently, methods that identify element oxidation states have been increasing. Schock et al. (2014b) used X-ray absorption near-edge spectroscopy to validate the identification by XRD of Pb(IV) compounds in pipe scales. X-ray absorption fine structure can also be used for valence state determination. These methods are likely to be used more in the future because of common ambiguities in the interpretation of XRD spectra in pipe scales (Schock and Lytle 2011).

When removing the pipes from the distribution system, bending the pipes should be avoided. Bending pipes will dislodge the scale. Pipe scales should remain moist once they are removed from the distribution system until they are cut at the lab. Dry scales are fragile and may dislodge during shipment to the laboratory. Sealing the pipe ends with plastic wrap will maintain the moisture (Figure 3-18). To prevent any damage, it is important that the pipe be well protected during delivery to the laboratory and that dropping and mishandling be avoided.

Courtesy of USEPA

Figure 3-18 Example of a capped pipe ready for transportation

Microbial Analyses

Several methods exist for analysis of microbial occurrence, which include the following:

- **Microscopy**—Used to assess the presence of microbes based on comparison of shape, size and color to known information. This method can be further refined by assessing the ability to stain the cell membrane or DNA with different dyes. DNA staining may show indication of cell viability.
- **Heterotrophic plate count**—Used to assess the presence of aerobic bacteria that use organic carbon as their carbon source. This test takes up to seven days for results and will only identify 0.1 percent to 1 percent of total population (Standard Method 9215, Heterotrophic Plate Count with R2A; APHA et al. 2012).
- **Most probable number**—A well-established and fully documented method of estimating the number of viable microorganisms in product in which the microorganisms are randomly distributed. This method offers real opportunities as a tool for microbiologists in situations where the plate count method is unsuitable (Standard Method 9222 for Coliform, Standard Method 9245 for Nitrifying Bacteria; APHA et al. 2012).
- **Polymerase chain reaction**—Used to assess the presence of specific bacteria, such as pathogens, or groups of bacteria of interest. Multiple tests are needed to assess the total population.
- **Adenosine triphosphate**—Used to assess total microbial occurrence based on the presence of ATP. The concentration of ATP can be converted to microbial equivalents based on an assumed amount of ATP per cell. Results can be obtained within 15 minutes of sample collection. Results can be biased by the presence of macroorganisms, which have more ATP per cell (ASTM D4012).

If a microbial analysis was performed and microbial occurrence was found to be excessive in comparison with other scale samples from the system, an analysis of conditions that support microbial growth should also be performed. This analysis could include physiochemical testing for changes in temperature, conductivity, DO concentration, ORP, pH, colored dissolved organic matter, and the concentration of nitrate in flowing water from a sample tap. This data could be used to identify the location of biofilm within a plumbing or distribution system and/or the presence of conditions different from the source water that could support microbial growth and potentially exacerbate corrosion reactions. Microbial presence and growth and corrosion are described earlier in this chapter in the "Biostability" section and also in chapter 2.

PREMISE PLUMBING PROFILES

In addition to pipe scale information, a utility may want to profile premise plumbing (in-house plumbing and service line) using wet chemistry data taken from sequential samples collected from the tap to the main (i.e., the cumulative volume of the sequential samples is equal to the volume of piping between the main and that tap) (Economic and Engineering Services and Illinois State Water Survey 1990). Premise plumbing profiles are useful to determine metal leaching from different portions of the residential plumbing, including LSLs, the stability of the pipe scales (i.e., how readily the scale is breaking off the pipe wall), the concentration and types of particulates dissociating from the scale, and microbiological activity with respect to nitrification and other localized MIC (Schock and Lytle 2011).

Currently, there is no standard procedure for conducting a premise plumbing profile. In some instances, a temperature change method is used, whereby sequential samples are collected until cold water reaches the faucet, indicating that freshwater from the service line has been reached. In other instances, a standard amount of samples is collected (12–25 or more individual liters), assuming that the entire plumbing profile will be captured. A more rigorous approach involves calculating the volume of water contained in a specific home's plumbing system based on knowledge of the site-specific plumbing configuration. Sandvig et al. (2008) reported premise plumbing volumes in field studies at multiple US utilities ranging from 0.9 to 5.4 L and service line volumes ranging from 1.1 to 11.1 L.

A recent study by Del Toral et al. (2013) used different sampling strategies to profile lead levels in homes with LSLs by comparing lead levels in standing 1-L and 45-second flushed samples, with and without pre-flushing prior to stagnation. The highest lead levels were consistently measured from the 45-second flushed samples, suggesting that current LCR monitoring protocols are underestimating lead release and human exposure.

Premise profiles should be conducted at several homes in the distribution system to establish a baseline (Economic and Engineering Services and Illinois State Water Survey 1990). At the end of the desired stagnation period (typically a minimum of 6 hours and a maximum of 24 hours), water is collected sequentially in 1-L bottles at the home until the calculated volume required to reach the main is met. Water should continue to run for several more minutes to obtain current water quality conditions coming from the main (Schock and Lytle 2011, Sandvig et al. 2008). Approximately 200 mL from each 1-L sample can be filtered to measure dissolved inorganic constituents. This filtering will allow the user to obtain both particulate and dissolved inorganic concentrations. Table 3-5 presents distance–volume relationships that have been developed for residential piping for investigating lead source.

To determine the physical stability of the scale, water hammer can be induced by repeatedly opening and closing the inside tap at full velocity, or higher velocity samples can be collected. Water hammer is a hydraulic pressure transient caused by rapid changes in flow rate caused by events such as pump station start-up and stops or power outages. Clearly, care must be taken to avoid damage to customer plumbing when conducting these types of tests. Samples can then be compared to the sequential samples just previously collected. Studies have consistently shown that moderate to high flow rates or flow disturbances can increase the mobilization of lead and result in significant contributions of particulate lead to the total lead concentration (Triantafyllidou et al. 2007; Deshommes et al. 2010; Cartier et al. 2011, 2012, 2013; Schock and Lytle 2011; Wang et al. 2012, 2013; Del Toral et al. 2013; Clark et al. 2014). Increased dissolved or particulate inorganic material such as lead, copper, manganese, phosphate (if a phosphate inhibitor is used), iron (from source water, upstream iron pipes, or galvanized material), and increased color may indicate an unstable scale formation and the potential for these constituents to leach or break off and find their way to the tap. Unlike unstable scales, water hammer samples with low inorganics or similar results to the sequential samples most likely indicate stable scales (Del Toral et al. 2013).

Table 3-5 Distance–volume relationships for some common residential piping, useful for planning, sampling, and diagnosing lead sources

Material	Type	Nominal Size (in.)	ID* (in.)	ID (mm)	mL/ft	mL/m	ft/L	m/L
Copper tube	K	½	0.527	13.4	43	141	23.3	7.1
	L	½	0.545	13.8	46	151	21.8	6.6
	M	½	0.569	14.5	50	164	20.0	6.1
	K	¾	0.745	18.9	86	281	11.7	3.6
	L	¾	0.785	19.9	95	312	10.5	3.2
	M	¾	0.811	20.6	102	333	9.8	3.0
	K	1	0.995	25.3	153	502	6.5	2.0
	L	1	1.025	26.0	162	532	6.2	1.9
	M	1	1.055	26.8	172	564	5.8	1.8
Galvanized steel†	Schedule 40	½	0.622	15.8	60	196	16.7	5.1
		¾	0.824	20.9	105	344	9.5	2.9
		1	1.049	26.6	170	558	5.9	1.8
Lead	¼-in. wall	½	0.500	12.7	39	127	25.9	7.9
		⅝	0.625	15.9	60	198	16.6	5.1
		¾	0.750	19.1	87	285	11.5	3.5
Polyvinyl chloride (PVC), chlorinated polyvinyl chloride (CPVC)‡	Schedule 80	½	0.546	13.9	46	151	21.7	6.6
		¾	0.742	18.8	85	279	11.8	3.6
		1	0.957	24.3	141	464	7.1	2.2
Polyethylene (HDPE)‡	200 psi	1	1.023	26.0	162	530	6.2	1.9

*ID = inside diameter.

†For old galvanized steel pipe, corrosion usually reduces the ID and volume considerably.

‡For plastic pipe, wall thickness can vary slightly, changing exact ID.

Source: *Schock and Lytle 2011*

Inorganic-based profile samples are usually analyzed for metals relevant to the water system, which may include lead, copper, iron, manganese, zinc, and aluminum. Other metals and inorganic anions (chlorides, sulfates, phosphates, etc.) may also be collected depending on the utility's treatment or specific situation.

In addition to inorganic scales, organic or microbiological scales may be assessed. Collecting sequential and water hammer samples for nitrites, AOC, total or free chlorine, ATP, and HPC-R2As may provide the utility with information regarding microbiological interactions that might increase localized corrosion. pH is also a recommended parameter to measure in premise profiles.

A utility experiencing localized premise copper corrosion from microbial activity may experience an increase in nitrite, a drop in free ammonia (if adding chloramine or in systems with naturally occurring ammonia), a decrease in pH in low-alkalinity water, an increase in copper, and a decrease in disinfectant residual in a premise plumbing profile after the 6-hour stagnation. Spikes in HPCs may occur during sampling; however, the spikes may not necessarily indicate corrosion is occurring or that corrosion is occurring at the particular location related to the sample number. For example, a spike might occur at the third liter, but that spike may not necessarily mean the biofilm was located at that

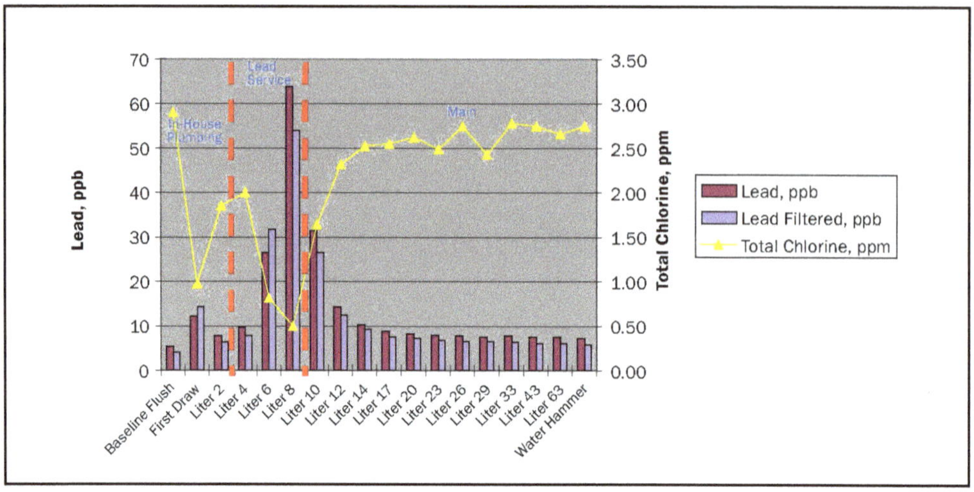

Courtesy of DC Water

Figure 3-19 Example of a lead profile at a residential home with a lead service line (lead = total lead; lead filtered = dissolved lead)

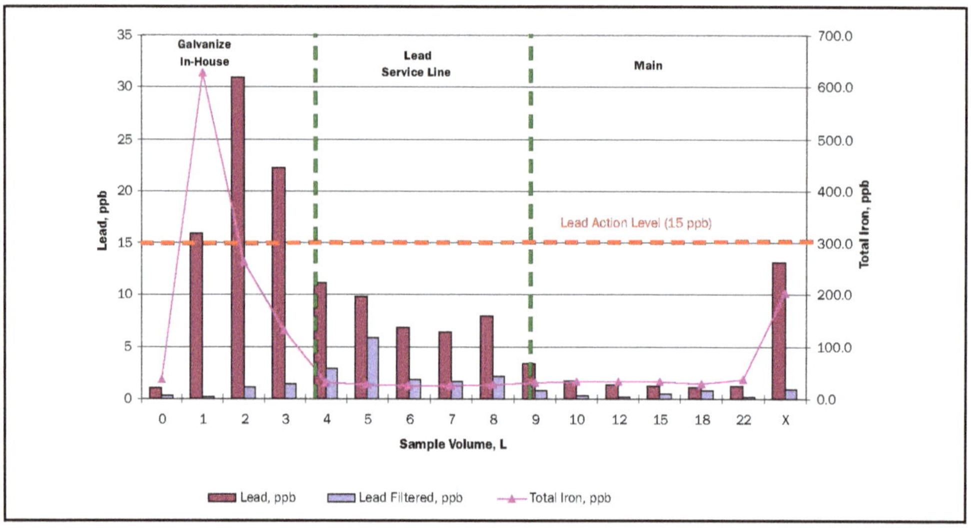

Source: *HDR Engineering 2009*

Figure 3-20 Example of a lead profile at a residential home with galvanized internal plumbing and lead service line

point. A spike from this sample location indicates that biofilm exists; however, the biofilm could be from the third sample location to the first sample location.

Biofilm may not always detach from the pipe wall immediately. In addition, absence of a spike in HPC does not always mean heterotrophic bacteria are not present. The flow may be too low in velocity to break up the biofilm, and therefore, HPC may never show up in the profile. Relying on the chemical parameters in conjunction with each other will probably provide more accurate information. However, if a biofilm exists, it will most likely be dislodged during the water hammer test.

Profiles in homes or buildings containing old galvanized lines where the zinc surface coating has dissolved will show elevated iron levels. In addition, elevated lead levels have been associated with this condition. Figures 3-19 and 3-20 graph residential profiles conducted at homes.

Once the samples are collected, graphing the analysis of each constituent might be helpful to quickly assimilate the data. Graphing total and dissolved metals is recommended. Figure 3-19 is an example of a lead graph with an overlay of total chlorine. Note how the chlorine dissipated over the stagnation period in the service line. In Figure 3-19, it is obvious that the main source of lead is from the service line in the form of a dissolved species (denoted as "filtered"), that is, lead(II).

SUMMARY

Part 1 of this chapter discussed water quality factors and other factors that influence corrosion and metals release, and provided the basis for a corrosion control monitoring program. Part 2 provided a framework by which utilities may assess and determine the cause of water quality problems from corrosion-related metals release.

Understanding the carbonate balance and its impact on corrosion and corrosion control treatment is critical to effective corrosion control. Other water quality parameters, such as oxidants, ORP, anions, co-occurring iron/manganese, and biostability, may also have a significant impact on metals release. In fact, MIC is often overlooked as a source of corrosion-related water quality problems, and recognition of the importance of biostability and its role in corrosion may help to control copper corrosion-related water quality problems that utilities experience. The importance of iron and manganese as metal scavengers and sinks for lead and other regulated metals is gaining more attention and becoming better understood.

Generally, the level of monitoring-required lead and copper regulations is insufficient for maintaining effective corrosion control treatment. More frequent monitoring of regulated parameters and other parameters at the source, entry point, and locations throughout the distribution system is recommended. It is also necessary to establish baseline water quality to effectively identify and implement corrosion control treatment and subsequently identify the cause of any potential corrosion-related water quality problems. Utilities may consider the use of compliance monitoring results to establish baseline water quality and effectiveness of the corrosion control treatment, and then develop a supplemental operational monitoring program to better manage and control the effectiveness of the program.

Lack of an adequate monitoring program will make it difficult to determine the cause of corrosion-related water quality problems. Identifying the cause of a corrosion-related water quality problem requires the evaluation of how three different categories of corrosion look and how they might possibly be linked to that problem.

ADDITIONAL READING

Readers may wish to review more information in the specific areas of

- the carbonate balance, more specifically pH, alkalinity, and DIC, and the role of these parameters in corrosion-related water quality problems and effective corrosion control;
- the role of other water quality factors on corrosion and the ability to effectively control metals release;
- the importance of biostability and the role of MIC in corrosion-related water quality problems; and
- accumulation of metals and other inorganics within the distribution system and premise plumbing scales and sediments.

References for this chapter give detailed information on these areas and the other topics discussed. Readers are encouraged to follow up on the referenced material that is publicly available to obtain keener insight into this discussion.

REFERENCES

APHA, AWWA, and WEF (American Public Health Association, American Water Works Association, and Water Environment Federation). 2012. *Standard Methods for the Examination of Water and Wastewater,* 22nd ed. Edited by E.W. Rice, R.B. Baird, A.D. Eaton, and L.S. Clesceri. Washington, D.C.: APHA, AWWA, and WEF.

ASTM D4012. 2015. *Standard Test Method for Adenosine Triphosphate (ATP) Content of Microorganisms in Water.* West Conshohocken, Pa.: ASTM International.

AWWA (American Water Works Association). 2005. *Managing Change and Unintended Consequences: Lead and Copper Rule Corrosion Control Treatment.* Denver, Colo.: AWWA.

AwwaRF and TZW (Awwa Research Foundation and DVGW-Technologiezentrum Wasser). 1996. *Internal Corrosion of Water Distribution Systems,* 2nd ed. Denver, Colo.: AwwaRF.

Benjamin, M.M., H. Sontheimer, and P. Leroy. 1996. Corrosion of iron and steel. In *Internal Corrosion of Water Distribution Systems,* 2nd ed. Denver, Colo.: AwwaRF and AWWA.

Bennett, P.C., J.R. Rodgers, and W.J Choi. 2001. Silicates, silicate weathering, and microbial ecology. *Geomicrobiology Journal,* 18:3–19.

Boissonneault, P. 1994. Dissolved inorganic carbon—A key to corrosion control in low alkalinity waters. In *Proceedings of the AWWA Water Quality Technology Conference.* Denver, Colo.: AWWA.

Boyd, G.R., P. Shetty, A.M. Sandvig, and G.L. Pierson. 2004. Pb in tap water following simulated partial lead pipe replacements. *Journal of Environmental Engineering,* 130(10):1188–1197.

Bremer, P.J., B.J. Webster, and D.B. Wells. 2001. Biocorrosion of copper in potable water. *Journal AWWA,* 93(8):82–91.

Broo, A.E., B. Berghult, and T. Hedburg. 1999. Drinking water distribution—The effect of natural organic matter (NOM) on the corrosion of iron and copper. *Water Science and Technology,* 40(9):17–24.

Burlingame, G.A., M.R. Schock, and M.A. Edwards. 2006. Knowing chemistry can help get the lead out. *Opflow,* 32(9):24–26.

Campbell, H.S., and M.E.D. Turner. 1983. The influence of trace organics on scale formation and corrosion. *Journal of the Institute for Water Engineering and Science,* 4:55.

Cantor, A.F. 2000. Use of polyphosphate in corrosion control. *Journal AWWA,* 92(2):95–102.

Cantor, A.F. 2006. Diagnosing corrosion problems through differentiation of metal fractions. *Journal AWWA,* 98(1):117.

Cantor, A.F., J.B. Bushman, and M.S. Glodoski. 2003. A new awareness of copper pipe failures in water distribution systems. In *Proceedings of the AWWA Water Quality Technology Conference*. Denver, Colo.: AWWA.

Cantor, A.F., J.B. Bushman, M.S. Glodoski, E. Kiefer, R. Bersch, and H. Wallenkamp. 2006. Copper pipe failure by microbiologically influenced corrosion. *Materials Performance*, (June):38–41.

Carlson, K., S. Via, B. Bellamy, and M. Carlson. 2000. Secondary effects of enhanced coagulation and softening. *Journal AWWA*, 92(6):63–75.

Cartier, C., L. Laroche, E. Deshommes, S. Nour, G. Richard, M. Edwards, and M. Prévost. 2011. Investigating dissolved lead at the tap using various sampling protocols. *Journal AWWA*, 103(3):53–67.

Cartier, C., R. Arnold, S. Triantafyllidou, M. Prévost, and M. Edwards. 2012. Effect of flow rate and lead/copper pipe sequence on lead release from service lines. *Water Research*, 46(13):4142–4152.

Cartier, C., E. Doré, L. Laroche, S. Nour, M. Edwards, and M. Prévost. 2013. Impact of treatment on Pb release from full and partially replaced lead service lines (LSLs). *Water Research*, 47(2):661–671.

Clark, B., S. Masters, and M. Edward. 2014. Profile sampling to characterize particulate lead risks in potable water. *Environmental Science and Technology*, 48(12):6836–6843. doi: 10.1021/es501342j

Clement, J.A., and M.R. Schock. 1998. Buffer intensity: What is it and why it's critical for controlling distribution system water quality. In *Proceedings of the AWWA Water Quality Technology Conference*. Denver, Colo.: AWWA.

Clifford, D.A. 1999. Ion exchange and inorganic adsorption. In *Water Quality and Treatment: A Handbook of Community Water Supplies*, 5th ed. Edited by R.D. Letterman. New York: McGraw-Hill.

Clifford, D., T. Sorg, and G. Ghurye. 2011. Ion exchange and adsorption of inorganic contaminants. In *Water Quality and Treatment: A Handbook on Drinking Water*, 6th ed. Edited by J.K. Edzwald. New York: McGraw-Hill.

Copeland, A., and D. Lytle. 2014. Measuring the oxidation-reduction potential of important oxidants in drinking water. *Journal AWWA*, 106(1):65–66.

De Haan, M., and M. Friedman. 2015. Removing accumulated manganese and metals from the distribution system. In *Proceedings of the AWWA Annual Conference and Exposition*. Denver Colo.: AWWA.

Del Toral, M., A. Porter, and M. Schock. 2013. Detection and evaluation of elevated lead release from service lines: A field study. *Environmental Science and Technology*, 47, 9300–9307.

Deshommes, E., L. Laroche, S. Nour, C. Cartier, and M. Prévost. 2010. Source and occurrence of particulate lead in tap water. *Water Research*, 44(12):3734–3744.

Dodrill, D.M., and M. Edwards. 1995. Corrosion control on the basis of utility experience. *Journal AWWA*, 87(7):74–85.

Economic and Engineering Services and Illinois State Water Survey. 1990. *Lead Control Strategies*. Denver, Colo.: AwwaRF and AWWA.

Edwards, M., and A. Dudi. 2004. Role of chlorine and chloramine in corrosion of lead-bearing plumbing materials. *Journal AWWA*, 96(10):69–81.

Edwards, M., and L.S. McNeill. 2002. Effect of phosphate inhibitors on lead release from pipes. *Journal AWWA*, 94(1):79–90.

Edwards, M., J.F. Ferguson, and S.H. Reiber. 1994a. The pitting corrosion of copper. *Journal AWWA*, 86(7):74–90.

Edwards, M., T. Meyer, and J. Rehring. 1994b. Effect of selected anions on copper corrosion rates. *Journal AWWA*, 86(12):73–81.

Edwards, M., S. Jacobs, and D. Dodrill. 1999. Desktop guidance for mitigating Pb and Cu corrosion by-products. *Journal AWWA*, 91(5):66–77.

Edzwald, J.K., and J.E. Tobiason. 2011. Chemicals principles, source water composition, and watershed protection. In *Water Quality and Treatment: A Handbook on Drinking Water*, 6th ed. Edited by J.K. Edzwald. New York: McGraw-Hill.

Escobar, I.C., and A.A. Randall. 2001. Case study: Ozonation and distribution system biostability. *Journal AWWA*, 93(10):77.

Federal Register. 1991. National Primary Drinking Water Regulations; Lead and Copper; Final Rule. *Federal Register*, 56(110):26460.

Ferguson, J.F., O. von Franque, and M.R. Schock. 1996. Corrosion of copper in potable water systems. In *Internal Corrosion of Water Distribution Systems*, 2nd ed. Denver, Colo.: AwwaRF and AWWA.

Fleming, K.K., G.W. Harrington, and D.R. Noguera. 2005. Nitrification potential curves: A new strategy for nitrification prevention. *Journal AWWA*, 97(8):90.

Friedman, M., and A. Blain. 2014. Use of monitoring, bench-scale, and pip rig studies to solve a pipe destabilization mystery. In *Proceedings of the AWWA Water Quality Technology Conference*. Denver, Colo.: AWWA.

Friedman, M.J., A.S. Hill, S.H. Reiber, R.L. Valentine, G.V. Korshin, C.Y. Peng, G. Larsen, and A. Young. 2010. *Assessment of Inorganics Accumulation in Drinking Water System Scales and Sediments*. Denver, Colo.: Water Research Foundation.

Friedman, M., A. Hill, S. Booth, M. Hallett, L. McNeill, J. McLean, D. Stevens, D. Dorensen, T. Hammer, W. Kent, M. De Haan, K. MacArthur, and K. Mitchell. 2016. *Metals Accumulation and Release Within the Distribution System: Evaluation and Mitigation*. Denver Colo.: Water Research Foundation.

HDR Engineering. 2009. *An Analysis of the Correlation Between Lead Released From Galvanized Iron Piping and the Contents of Lead in Drinking Water.* Bellevue, Wash.: HDR Engineering. Available at: www.dcwater.com/waterquality/plumbing/galvanized.cfm.

Hill, A.S., M.J. Friedman, S.H. Reiber, G.V. Korshin, and R.L. Valentine. 2010. Behavior of trace inorganic contaminants in drinking water distribution systems. *Journal AWWA*, 102(7):107–118.

Hill, C. 2016. Lead and Copper Rule: Potential Revisions and Steps Utilities Can Take to Prepare. AWWA Webinar 160525, Slides 55–74. May 25, 2016. Available at: www.awwa.org/Portals/0/files/education/webcast/W160525Handouts.pdf

Horsley, M.B., B.W. Northup, W.J. O'Brien, and L.L. Harms. 1998. Minimizing iron corrosion in lime softened water. In *Proceedings of the AWWA Water Quality Technology Conference*. Denver, Colo.: AWWA.

Imran, S.A., J.D. Dietz, G. Mutoti, J.S. Taylor, A.A. Randall, and C.D. Cooper. 2005. Red water release in drinking water distribution systems. *Journal AWWA*, 97(9):93–100.

Ives, D.J., and R.W. Rawson. 1962. Copper corrosion IV: The effects of saline additions. *Journal of the Electrochemical Society*, 109(6):462.

Jacobs, S., S. Reiber, and M. Edwards. 1998. Sulfide-induced copper corrosion. *Journal AWWA*, 90(7):62–73.

Kimbrough, D.E. 2007. Brass corrosion as a source of lead and copper in traditional and all-plastic distribution systems. *Journal AWWA*, 99(8):70–76.

Kirchner, G. 2009. Clintonville School water OK to drink. *Shawano Leader*, Feb. 25, 2009.

Korshin, G.V., J.F. Ferguson, and A.N. Lancaster. 2005. Influence of natural organic matter on the morphology of corroding lead surfaces and the behavior of lead-containing particles. *Water Research*, 39(5):811–818.

Larson, T.E. 1966. Chemical control of corrosion. *Journal AWWA*, 58(3):354.

Lauer, W.C., and S.R. Lohman. 1994. Non-calcium carbonate protective film lowers lead levels. In *Proceedings of the AWWA Water Quality Technology Conference*. Denver, Colo.: AWWA.

Lin, Y-P., and R.L. Valentine. 2008. Release of Pb(II) from monochloramine-mediated reduction of lead oxide (PbO_2). *Environmental Science and Technology*, 42(24):9137–9143.

Lowry, J. 2009. Lakhurst Acres, ME: Compliance Issues Engineering Problems and Solutions. US EPA Sixth Annual Drinking Water Workshop: Small Drinking Water System Challenges and Solutions, Cincinnati, Ohio, Aug. 4–6, 2009.

Lowry, J. 2010. Corrosion Control With Air Stripping. Inorganic Contaminants Workshop. American Water Works Association, Denver, Colo., Feb. 28–Mar. 2, 2010.

Lytle, D.A., and M.R. Schock. 2005. Formation of Pb(IV) oxides in chlorinated water. *Journal AWWA*, 97(11):102–114.

Lytle, D.A., P. Sarin, and V.L. Snoeyink. 2003. The effect of chloride and orthophosphate on the release of iron from drinking water distribution system cast iron pipe. In *Proceedings of the AWWA Water Quality Technology Conference*. Denver, Colo.: AWWA.

Lytle, D.A., J.M. Payne, and M. Schock. 2004. Studying distribution system solids: Approach and application. In *Proceedings of the AWWA Annual Conference and Exposition*. Denver, Colo.: AWWA.

Maynard, B., and D. Mast. 2006. Composition of interior scales on lead source materials. In *Proceedings of the AWWA Water Quality Technology Conference*. Denver, Colo.: AWWA.

McFadden, M., R. Giani, P. Kwan, and S.H. Reiber. 2011. Contributions to drinking water lead from galvanized iron corrosion scales. *Journal AWWA*, 103(4):76.

McNeill, L.S., and M. Edwards. 2001. Iron pipe corrosion in distribution systems. *Journal AWWA*, 93(7):88–100.

Nguyen, C., K. Stone, B. Clark, M. Edwards, G. Gagnon, and A. Knowles. 2010. *Impacts of Chloride:Sulfate Mass Ratio (CSMR) Changes on Lead Leaching in Potable Water*. Denver, Colo.: Water Research Foundation.

Obrecht, M.F., and L.L. Quill. 1960. How temperature, treatment, and velocity of potable water affect corrosion of copper and its alloys; tests show effects of water quality at various temperatures, velocities. *Heating, Piping, and Air Conditioning*, 32(5):105.

Obrecht, M.F., and L.L. Quill. 1961. How temperature, velocity of potable water affect corrosion of copper and its alloys. *Heating, Piping, and Air Conditioning*, 33(4):129.

Pisigan, R.A., and J.E. Singley. 1987. Influence of buffer capacity, chlorine residual, and flow rate on corrosion of mild steel and copper. *Journal AWWA*, 79(2):62–70.

Post, G.B., T.B. Atherholt, and P.D. Cohn. 2011. Health and aesthetic aspects of drinking water. In *Water Quality and Treatment: A Handbook on Drinking Water*, 6th ed. Edited by J.K. Edzwald. New York: McGraw-Hill.

Reiber, S. 2004. Disinfection Byproducts vs. Corrosion: A Case Study on the DC WASA Lead Experience. Presentation to USEPA Technical Expert Working Group, Washington, D.C., April 2004.

Reiber, S. 2006. Corrosion, red water, and distribution system water quality. In *Proceedings of the AWWA Annual Conference and Exposition*. Denver, Colo.: AWWA.

Reiber, R., and G. Dostal. 2000. Well water disinfection sparks surprises. *Opflow*, 26(3):1, 4–6, 14.

Reiber, S.H., J.F. Ferguson, and M.M. Benjamin. 1987. Corrosion monitoring and control in the Pacific Northwest. *Journal AWWA*, 79(2):71–74.

Renner, R. 2006. Experiment confirms chloramine's effect on lead in drinking water. *Environmental Science and Technology*, 40(10):3129–3130.

Rogers, J.R., and P.C. Bennett. 2004. Mineral stimulation of subsurface microorganisms: Release of limiting nutrients from silicates. *Chemical Geology*, 203:91–108.

Rushing, J.C., and M. Edwards. 2002. Effect of aluminum solids and chlorine on cold water pitting of copper. In *Proceedings of the AWWA Water Quality Technology Conference*. Denver, Colo.: AWWA.

Sandvig, A., P. Kwan, G. Kirmeyer, B. Maynard, D. Mast, R. Rhodes Trussell, S. Trussell, A. Cantor, and A. Prescott. 2008. *Contribution of Service Line and Plumbing Fixtures to Lead and Copper Rule Compliance Issues*. Denver, Colo.: Water Research Foundation.

Sarin, P., J. Bebee, M.A. Beckett, K.K. Jim, D.A. Lytle, J.A. Clement, W.M. Kriven, and V.L. Snoeyink. 2000. Release of Iron From Corroded Iron/Steel Pipes in Water Distribution Systems—Effect of Stagnation Time. International Distribution Research Symposium, Denver, Colo., June 10–11, 2000.

Sarin, P., J.A. Clement, and V.L. Snoeyink. 2003. Iron release from corroded unlined cast-iron pipe. *Journal AWWA*, 95(11):85–96.

Scanlan, L. 2003. Distribution system water quality changes: A challenge. In *Proceedings of the AWWA Annual Conference and Exposition*. Denver, Colo.: AWWA.

Scardina, P.W. 2006. Copper pinhole leaks. In *Proceedings of the AWWA Annual Conference and Exposition*. Denver, Colo.: AWWA.

Schock, M.R. 1999. Internal corrosion and deposition control. In *Water Quality and Treatment: A Handbook of Community Water Supplies*, 5th ed. Edited by R.D. Letterman. New York: McGraw-Hill.

Schock, M.R. 2005. Distribution systems as reservoirs and reactors for inorganic contaminants. In *Distribution System Water Quality Challenges in the 21st Century—A Strategic Guide*. Edited by M.J. McPhee. Denver, Colo.: AwwaRF.

Schock, M., and M. DeSantis. 2014. Ground Truthing the "Conventional Wisdom" of Lead Corrosion Control Using Mineralogical Analysis. Presented at AWWA Water Quality Technology Conference, New Orleans, La., Nov. 16–19, 2014.

Schock, M.R., and R. Giani. 2005. Oxidant/Disinfectant chemistry and impacts on lead corrosion. In *Proceedings of the AWWA Water Quality Technology Conference*. Denver, Colo.: AWWA.

Schock, M.R., and D.A. Lytle. 1995. Control of copper corrosion of household plumbing materials. In *Abstract, Proceedings of the 21st Annual USEPA RREL Research Symposium*. Cincinnati, Ohio: USEPA.

Schock, M., and D. Lytle. 2011. Internal corrosion and deposition control. In *Water Quality and Treatment: A Handbook on Drinking Water*, 6th ed. Edited by J.K. Edzwald. New York: McGraw-Hill.

Schock, M.R., D.A. Lytle, and J.A. Clement. 1995. *Effect of pH, DIC, Orthophosphate and Sulfate on Drinking Water Cuprosolvency*. EPA/600/R-95/085. Cincinnati, Ohio: USEPA, Office of Research and Development.

Schock, M.R., I. Wagner, and R.J. Oliphant. 1996. Corrosion and solubility of lead in drinking water. In *Internal Corrosion of Water Distribution Systems*, 2nd ed. Denver, Colo.: AwwaRF and AWWA.

Schock, M.R., J. Clement, and D.A. Lytle. 1998. Replacing polyphosphate with silicate to solve problems with lead, copper and source water iron. In *Proceedings of the AWWA Water Quality Technology Conference*. Denver, Colo.: AWWA.

Schock, M.R., T.L. Geske, M.K. DeSantis, and R.C. Copeland. 2006. *Scale Analysis for Madison, WI, Samples*. Cincinnati, Ohio: USEPA.

Schock, M.R., R.N. Hyland, and M.M. Welch. 2008. Occurrence of contaminant accumulation in lead pipe scales from domestic drinking water distribution systems. *Environmental Science and Technology*, 42(12):4285.

Schock, M., F. Lemieux, and N. Martinova. 2014a. Framework for Developing pH Guidance for Drinking Water Treatment and Distribution. Presented at 16th Annual Canadian National Conference on Drinking Water, Gatineau, QC, Canada, Oct. 26–29, 2014.

Schock, M.R., A.F. Cantor, S. Triantafyllidou, M.K. Desantis, and K.G. Scheckel. 2014b. Importance of pipe scales to Lead and Copper Rule compliance. *Journal AWWA*, 106(7):E336–E349.

Smith, C.D. 2006. Monitoring for nitrification prevention and control. In Manual M46. *Fundamentals and Control of Nitrification in Chloraminated Drinking Water Distribution Systems*. Denver, Colo.: AWWA.

Smith, S.E., T. Ta, D.M. Holt, A. Delanoue, and J. Colbourne. 1998. Minimising red water in drinking water distribution systems. In *Proceedings of the AWWA Water Quality Technology Conference*. Denver, Colo.: AWWA.

Snoeyink, V.L., and D. Jenkins. 1980. *Water Chemistry*. New York: John Wiley and Sons.

Snoeyink, V.L., and I. Wagner. 1996. Principles of corrosion of water distribution systems. In *Internal Corrosion of Water Distribution Systems*, 2nd ed. Denver, Colo.: AwwaRF and AWWA.

Sontheimer, H., W. Kolle, and R. Rudek. 1979. Aufgaben und Methoden der Wasserchemie—Dargestellt an der Entwicklung erkenntnisse zur Bildung von Korrosions-shutzschichten auf Metallen. *Vom Wasser*, 52:1.

Sontheimer, H., W. Kolle, and V.L. Snoeyink. 1981. The siderite model of the formation of corrosion-resistant scales. *Journal AWWA*, 73(11):572.

Sorg, T.J., M.R. Schock, and D.A. Lytle. 1999. Ion exchange softening: Effects on metals concentrations. *Journal AWWA*, 91(8):85–97.

Stone, A., D. Spyridakis, M. Benjamin, J. Ferguson, S. Reiber, and S. Osterhus. 1987. The effects of short-term changes in water quality on copper and zinc corrosion rates. *Journal AWWA*, 79(2):75–82.

Stumm, W. 1960. Investigation of the corrosive behavior of waters. *ASCE Journal of the Sanitary Engineering Division*, 86(SA6):27.

Stumm, W., and J.J. Morgan. 1996. *Aquatic Chemistry: Chemical Equilibria and Rates in Natural Water*, 3rd ed. New York: John Wiley and Sons.

Triantafyllidou, S. 2006. Addressing and Assessing Lead Threats in Drinking Water: Non-Leaded Brass, Product Testing, Particulate Lead Occurrence and Effects of the Chloride to Sulfate Mass Ratio on Corrosion. Master's thesis. Virginia Polytechnic Institute and State University.

Triantafyllidou, S., and M. Edwards. 2007. Role of chloride to sulfate mass ratio in lead leaching from soldered joints and brass. *Journal AWWA*, 99(7):96–109.

Triantafyllidou, S., J. Parks, and M. Edwards. 2007. Lead particles in potable water. *Journal AWWA*, 99(6):107–117.

USEPA (US Environmental Protection Agency). 1994. *Method 200.8: Determination of Trace Elements in Waters and Wastes by ICP-MS*. Revision 5.4. Cincinnati, Ohio: USEPA, Environmental Monitoring Systems Laboratory, Office of Research and Development.

USEPA. 1996. *Method 3050B: Acid Digestion of Sediments, Sludges, and Soils*. Revision 2. CD-ROM. Cincinnati, Ohio: USEPA. Available at: https://www.epa.gov/homeland-security-research/epa-method-3050b-acid-digestion-sediments-sludges-and-soils

USEPA. 2006. *Inorganic Contaminant Accumulation in Potable Water Distribution Systems*. Total Coliform Rule Issue Paper. Washington, D.C.: USEPA, Office of Water.

USEPA. 2007. *Elevated Lead in DC Drinking Water*. EPA 815-R-07-021. Washington, D.C.: USEPA, Office of Water.

USEPA. 2012. Radionuclides in drinking water, compliance options: Treatment technology descriptions. Available at: http://cfpub.epa.gov/safewater/radionuclides/radionuclides.cfm

USEPA. 2016. *Corrosion Control Treatment Evaluation Technical Recommendations for Primacy Agencies and Public Water Systems*. EPA 816-B-16-003. Washington, D.C.: USEPA, Office of Water.

Van der Kooij, D. 1992. Assimilable organic carbon as an indicator of bacterial regrowth. *Journal AWWA*, 84(2):57.

Van der Merwe, S.W. 1988. The effect of water quality variables on the corrosive behaviour of water coagulated with a cationic polyelectrolye and with lime/activated silica. *Water Supply*, 6(4):SS2.

Wang, L., A.S.C. Chen, and A. Wang. 2010. *Arsenic Removal From Drinking Water by Ion Exchange: US EPA Demonstration Project at Fruitland, Idaho—Final Performance Evaluation Report*. EPA/600/R-10/152. Cincinnati, Ohio: USEPA.

Wang, Y., H. Jing, V. Mehta, G.J. Welter, and D.E. Giammar. 2012. Impact of galvanic corrosion on lead release from aged lead service lines. *Water Research*, 46:5049–5060.

Wang, Y., V. Mehta, G.J. Welter, and D.E. Giammar. 2013. Effect of connection methods on lead release from galvanic corrosion. *Journal AWWA*, 105(7):E337–E351.

Wolfe-Simon, F., J.S. Blum, T.R. Kulp, G.W. Gordon, S.E. Hoeft, J. Pett-Ridge, J.F. Stolz, S.M. Webb, P.K. Weber, P.C.W. Davies, A.D. Anbar, and R.S. Oremland. 2011. A bacterium that can grow by using arsenic instead of phosphorus. *Science*, 332(3):1163–1166.

Wu, L., A.D. Jacobsen, and M. Hausner. 2008. Characterization of elemental release during microbe-granite interactions at T=28°C. *Geochemica et Cosmochimica Acta*, 72:1076–1095.

Zhang, M., M.J. Semmens, D. Schuler, and R.M. Hozalski. 2002. Biostability and microbiological quality in a chloraminated distribution system. *Journal AWWA*, 94(9):112.

Chapter 4

Corrosion Control Treatment Techniques

Rebecca Slabaugh, Arcadis
Christopher P. Hill, Arcadis
Elizabeth Turner, North Texas Municipal Water District
Richard Giani, CH2M

INTRODUCTION

The implementation of corrosion control can be accomplished in eight steps.

1. Develop an understanding of factors affecting internal corrosion.
2. Determine the extent and magnitude of corrosion.
3. Determine the possible causes of corrosion.
4. Assess corrosion control alternatives.
5. Select a corrosion control strategy.
6. Implement a corrosion control program.
7. Monitor the effectiveness of the corrosion control program.
8. Optimize the control program if necessary.

Selection of a corrosion control technique may not be a one-time event. Any change in water treatment or source water may require a system to reevaluate corrosion control by following the preceding steps listed. Earlier chapters discussed corrosion theory and how to determine possible causes of corrosion. This chapter covers the assessment of corrosion control alternatives and some tools to help select a corrosion control strategy (Table 4-1). The complete elimination of corrosion is almost impossible. Technologies exist to reduce

Table 4-1 Chapter 4 key points

- Chemical treatment options include pH adjustment, dissolved inorganic carbon adjustment, and application of corrosion inhibitors.
- Techniques for controlling microbiologically influenced corrosion include disinfection, control of water age, flushing, and nutrient control.
- Utilities may find it necessary to reoptimize corrosion control treatment whenever changes are encountered in the source water or when utilities are contemplating treatment changes such as changes of coagulant type or secondary disinfectant.
- Implementation or reoptimization of corrosion control may have secondary impacts, such as changes in precipitation of solids, reduced disinfection efficacy, microbial growth, and disinfectant by-product formation, and these potential impacts must be considered when selecting a corrosion control method.
- Bench tests can be used to help select a treatment method and identify potential impacts that may result from adding a new chemical or altering corrosion control treatment practices.

or inhibit corrosion, but the feasibility of each technology will vary from system to system. The success of a particular corrosion control treatment is dependent on the specific water quality and piping of an individual system.

Implementing and maintaining optimum corrosion control treatment require balancing the needs of conflicting water quality goals. For example, the optimum pH for coagulation is not the optimum pH for orthophosphate used for corrosion control.

CORROSION CONTROL TREATMENT

Corrosion control is achieved through *passivation*, where pipe material and existing scale deposits are used to form a barrier film of less soluble metal carbonates or phosphate compounds on the inner pipe surface. This barrier layer of metal complex isolates the piping materials from the water supply and minimizes dissolution.

Treatment options include (1) pH and/or alkalinity adjustment and (2) application of corrosion inhibitors. More than one viable treatment option may be available to a water system. The tools presented in this chapter will help the reader select the best treatment option. Both Health Canada's *Guidance on Controlling Corrosion in Drinking Water Distribution Systems* (2009) and the World Health Organization's *Guidelines for Drinking-Water Quality* (WHO 2011) discuss common strategies for various metals including copper, iron, and lead. Additionally, USEPA's *Optimal Corrosion Control Treatment Evaluation Technical Recommendations for Primacy Agencies and Public Water Systems* (2016) includes several sets of flow charts, which can be used to identify viable treatment options based on water quality characteristics. Table 4-2 summarizes various corrosion control treatment methods and presents different water quality and operational advantages and disadvantages associated with each treatment method. A list of common corrosion control treatment chemicals is provided in Table 4-3.

pH and/or Alkalinity Adjustment

The adjustment of pH is a common method for reducing corrosion in distribution systems. The water pH is the major factor that determines the solubility of most pipe materials and the films that form from corrosion by-products.

The relationship of pH, dissolved inorganic carbon (DIC), alkalinity, carbon dioxide, and ionic strength governs the solubility of metal carbonates, which are commonly involved in protective scale formation on interior pipe surfaces. The type of alkalinity present and the pH of the water are dependent on each other.

Table 4-2 Advantages and disadvantages of various corrosion control treatment methods

Treatment Method	Advantages	Disadvantages
	pH and/or Alkalinity Adjustment	
Caustic soda/ caustic potash	• Adds hydroxide alkalinity. • Small capital and investment cost. • Easy to dose.	• Overdose in low-alkalinity water with low carbon dioxide (CO_2) increases pH to excessively high values. • Hazardous chemical; severe burns can occur to skin or eyes on exposure. • Caustic soda has a high freezing point; 50% caustic can freeze or crystallize at 58°F (14°C).
Soda ash/potash	• Increases total alkalinity by 0.94 mg/L as $CaCO_3$ per 1 mg/L soda ash or potash added. • Increases dissolved inorganic carbon (DIC) and pH to above 8.3. • Mainly used in powdered form, delivered in bulk or in sacks. • Nonhazardous chemical. • Relatively easy to handle and to dose.	• Adds 0.43 mg/L sodium per 1 mg/L soda ash added. • Hydrolyzes by absorbing water from air, which can make it cake; sacks should not be stored more than 6 months. • Higher cost than that of lime (roughly 10 times). • Requires pH control. • Requires dust control.
Sodium bicarbonate	• Used primarily to increase alkalinity with minimal increase in pH. • Alkalinity increases by 0.59 mg/L as $CaCO_3$ for 1 mg/L dose. • Nonhazardous chemical. • Relatively easy to handle and dose.	• Cannot raise pH > 8.3. • Higher cost than other chemicals (e.g., about 40 times higher than calcium carbonate). • Requires pH control. • Requires dust control.
Hydrated lime	• Increases pH and calcium (hardness) for soft water. • Adds hydroxide alkalinity. • Relatively inexpensive chemical.	• Overdose in soft waters with low free CO_2 will increase pH above the acceptable range; therefore, addition of lime to soft waters is sometimes accompanied by CO_2. • Varies in quality and may contain contaminants, such as grit and sand, heavy metals, and high concentration of aluminum. If high dose of lime is needed, the increase in Al concentration can be as high as 0.5 mg/L. The contribution of Al from lime can be minimized by adding it during the coagulation process. • Often increases turbidity of treated water. • May clog dosing equipment. • Easier to store than quicklime, but carbonation can cause deterioration. • Low winter temperatures in an unheated building could lead to freezing or caking. • Requires large investment in chemical dosing equipment. • Need to have sedimentation tank or saturator to prevent lime particles from going into the treated water. In other cases, deep bed filtration is used after lime addition. • May cause dust problems if special precautions are not taken. • Turbidity and pH control are necessary; pH control is difficult for low-alkalinity water.

Table continues next page

Table 4-2 Advantages and disadvantages of various corrosion control treatment methods (continued)

Treatment Method	Advantages	Disadvantages
Quicklime	• Increases pH and calcium (hardness) for soft water. • Adds hydroxide alkalinity. • Often slightly less expensive than hydrated lime and is more cost-effective when used in very large quantities.	• Overdose of lime in soft waters with low free CO_2 will increase pH above the acceptable range. Therefore, addition of lime to soft waters is sometimes accompanied by CO_2. • Varies in quality and may contain contaminants, such as grit and sand, heavy metals, and high concentration of aluminum. If high dose of lime is needed, the increase in Al concentration can be as high as 0.5 mg/L. The contribution of Al from lime can be minimized by adding it during the coagulation process. • Often increases turbidity of treated water. • Requires a lime slaker, which is a piece of mechanical equipment designed to control the reaction between quicklime and water to produce a hydrated lime slurry. Quicklime is difficult to use because of its low solubility and high hygroscopy (or ability to attract and hold water), which increase the risks of caking and blockages. • Must be stored in moisture-proof areas and used within a few weeks of manufacture to avoid absorbing water and CO_2 and forming calcium carbonate cake. • May cause dust problems if special precautions are not taken.
Limestone contactor	• Increases hardness, pH, and alkalinity of soft water. • Eliminates the need to feed chemicals and cannot result in overdosing. Claimed to be low cost and simple operation.	• Addition of CO_2 is often necessary to prevent excessive pH values in low-alkalinity water treated with half-burned dolomite ($CaMg(CO_3)_2$) filters. CO_2 gas is sometimes used to address this issue. • Varies in quality and may contain numerous impurities: clay, silt and sand, iron, phosphorus, sulfur, and trace amounts of manganese, copper, sodium, and arsenic (USEPA 1986). • No control over pH, which depends on the flow rate and could vary more than 1 unit. The rate of dissolution of $CaCO_3$ may vary depending on impurities in limestone (USEPA 1986). • Equilibration of the column with atmospheric CO_2 can have a significant impact on pH (USEPA 1986). • Fairly large investment in an alkaline media filter or a limestone contactor. The dissolution rate for limestone is low. Half-burned dolomite has a much higher dissolution rate and is, therefore, preferred in many cases. Requires a better pH control system. Very high pH values can occur. • Needs to be backwashed when clogging reduces the flow capacity.
Aeration	• Removes CO_2, which increases the pH and decreases DIC.	• Final pH is dependent on the amount of CO_2 in the source water.
Carbon dioxide	• Increases alkalinity and DIC if added prior to a base chemical. • Typically added prior to caustic soda in extremely low-alkalinity water to provide adequate alkalinity in the treated water.	• Decreases pH and requires additional base chemical. • Increases $CaCO_3$ precipitation potential.

Table continues next page

Table 4-2 Advantages and disadvantages of various corrosion control treatment methods (continued)

Treatment Method	Advantages	Disadvantages
	Corrosion Inhibitors	
Orthophosphate products	• Effective at moderate pH 7.0–8.0 for copper and lead control. Orthophosphate can reduce lead solubility by building up a lead phosphate layer that has a low solubility. Reduced copper corrosion rate in cold waters. Effectively reduces the general corrosion of the zinc layer of galvanized steel. Can reduce corrosion rate of steel and impact iron-release rate. • Normally have an inhibiting effect if the corrosion products or the corroding metal can form insoluble phosphates and improve the protective scale by increasing its impermeability and adherence. • Various blends and manufacturers available.	• Insufficient dose may accelerate corrosion or may not avoid pitting or other localized corrosion. Very pH-dependent with copper corrosion protection lost at slightly acidic pH values. • Higher phosphate levels (>3.0 mg/L as PO_4) can cause milky water in the distribution system with waters containing high calcium and aluminum in the water or on the pipe scales (>0.1 mg/L) (Tesfai et al. 2006, Frommel et al. 2004). • Orthophosphate can stimulate algal growth in receiving waters. • Ineffective for copper control outside pH range of 7.0–8.0 (Reiber et al. 1997). • Corrosion control may require several months of dosing at high initial doses before corrosion control benefits are fully observed. • The protective films will gradually break down and protection will cease within a short time if the dosage is stopped. • Pilot studies are often required because the level of corrosion reduction can vary substantially with inhibitor.
Zinc orthophosphate	• Corrosion control of lead, copper, iron, and asbestos cement under select water quality conditions.	• Many utilities use a combination of zinc sulfate and phosphoric acid. The mechanism of zinc action on lead and copper surfaces has never been quantified. Zinc orthophosphate has not offered significant benefit versus simple orthophosphates. • Zinc is currently not allowed in corrosion inhibitor products in much of Europe and is becoming increasingly restricted in North America because of zinc discharge limits in receiving waters. • Corrosion control may require several months of dosing before corrosion control benefits are fully observed. • The protective films will gradually break down and protection will cease within a short time if the dosage is stopped. • Pilot studies are often required because the level of corrosion reduction can vary substantially with inhibitor.
Polyphosphate products	• Prevent red water for iron corrosion. • Breakdown of polyphosphate to orthophosphate in distribution system has some benefit on lead and copper corrosion. • Maximum benefit for lead control will occur in the lowest alkalinity waters (Reiber et al. 1997). • Sequesters iron, manganese, and calcium in solution. • Various blends and manufacturers available.	• Unknown how and why they work. May increase zinc, copper, and lead corrosion. Polyphosphates may increase lead levels by complexation. Usually have little or no effect on either corrosion inhibition or the reduction of metals release. • Adverse effects on lead corrosion control for waters with alkalinity between 30 and 74 mg/L $CaCO_3$. Variable results for copper control above pH 7.8 (Reiber et al. 1997). • Polyphosphates are capable of solubilizing metal oxides and dissolving some corrosion scales and likely not suitable for lead and copper control (Reiber et al. 1997).

Table continues next page

Table 4-2 Advantages and disadvantages of various corrosion control treatment methods (continued)

Treatment Method	Advantages	Disadvantages
Blended phosphates	• Dose is dependent on pH and Ca concentration. Low doses are used to avoid soft scale formation and discolored water. • Phosphates form protective films on the bare metal surface by chemical reaction with either the corrosion products or the bare metal. • Various proprietary blends, typically 30%–70% orthophosphate, and manufacturers available.	• Insufficient doses can aggravate pitting corrosion. Blends with small proportion of orthophosphates can increase corrosion rates and iron release. • Variable corrosion inhibition has been found for mixtures of orthophosphates and polyphosphates. • Proprietary phosphate blends are typically more expensive than generic orthophosphate and may not provide additional corrosion control advantages. • Polyphosphate inhibitors had adverse effects on lead corrosion above pH 7.0, likely a result of complexation and release of lead corrosion by-products (Reiber et al. 1997). • Corrosion control may require several months of dosing before corrosion control benefits are fully observed. • The protective films will gradually break down and protection will cease within a short time if the dosage is stopped. • Pilot studies are often required because the level of corrosion reduction can vary substantially with an inhibitor.
Silicates	• Typically applicable for soft waters with low pH. • A major benefit of using silicates appears to be the accompanying pH increase. It is often more economical to increase the pH to between 7.5 and 8.0 by adding caustic soda or soda ash to lower the silicate requirement. The corrosion protection appears to be caused mainly by the pH increase and film formation. May reduce red-water problems, more so than pH or pH and alkalinity adjustment alone. • Protection by silicates occurs by forming a thin layer over the corroded metal layer. The films do not build up and are self-limiting.	• Poorly soluble in cold water. • Corrosion must occur before the metal surface can be protected due to mechanism for inhibition being the film of ferric or another metal oxide and silicate. • The protective films will gradually break down and protection will cease within a short time if the dosage is stopped. • Generally less effective than orthophosphates. Often observed effects depend on the pH increase caused by the alkaline character of the silicate compound. • Silicates do not inhibit zinc corrosion beyond the effect of increased pH. • Silicates can stimulate certain biological growth in low-mineralized waters. • Silicate component of caustic silicates (apart from pH adjustment) show little evidence of any inhibitory impact on either lead or copper. • Limited evidence suggests that silicate addition without pH control may exacerbate copper corrosion (Reiber et al. 1997). • Typically higher cost than phosphates and bimetallic inhibitors. • Pilot studies are often required because the level of corrosion reduction can vary substantially with an inhibitor.
Phosphate–silicate blends	• Provide corrosion protection as good as phosphate alone but at reduced phosphate concentrations. Dual mechanism may provide better corrosion protection. • Phosphate–silicate blends often show good corrosion reduction in steel pipe.	• Even in the case of very high pH of 8.7, when the iron release rate decreases to very low values, the corrosion rate remains relatively high in the low-mineralized water. The result of this high corrosion rate and low iron-release rate is that a large quantity of ferric precipitates is deposited in the pipes. • Pilot studies are often required because the level of corrosion reduction can vary substantially with an inhibitor.

Table 4-3 Characteristics of common corrosion control treatment chemicals

Chemical Name and Formula	Common Name	Common Forms	Approximate Commercial Strength
Calcium carbonate, $CaCO_3$	Limestone	White or gray; granular or powder	95%
Calcium hydroxide, $Ca(OH)_2$	Hydrated lime, slaked lime	White powder	72%–74% as CaO
Calcium oxide, CaO	Burnt lime, chemical lime, quicklime, unslaked lime	White; lump, pebble, granular, or pulverized	93%–99% CaO
Carbon dioxide, liquid CO_2	Carbonic anhydride	Liquid	99.5%
Phosphoric acid, H_3PO_4	—	Watery, white liquid	35%–85%
Potassium carbonate, K_2CO_3	Potash	White powder	>98.5%
Potassium hydroxide, KOH	Caustic potash	Clear liquid	45% or 50%
Sodium bicarbonate, $NaHCO_3$	Baking soda	White powder	99%
Sodium carbonate, Na_2CO_3	Soda ash	White powder	>98%
Sodium hexametaphosphate, $Na(PO_3)_6$	—	Crystal, flake, or powder	66%
Sodium hydroxide, NaOH	Caustic soda, soda lye	White solid beads, flakes, or liquid	98.9% (NaOH solid) 76% (Na_2O) 25% or 50% (NaOH solution)
Sodium silicate, Na_2SiO_3	Waterglass	Opaque, viscous liquid	38–42%
Zinc orthophosphate, $Zn_3(PO_4)_2$	—	White or clear liquid	Varies by manufacturer

Adapted from AwwaRF and TZW 1996, AWWA 2004

The solubility of metals typically decreases as pH increases (Dodrill and Edwards 1995, Reiber et al. 1997). Increased pH may, however, cause excessive calcium carbonate precipitation in waters with high hardness, resulting in hydraulic issues and increased pumping costs. Increasing pH can also result in higher trihalomethane (THM) concentrations. Additionally, iron and manganese, if present at elevated concentrations, may precipitate out and cause aesthetic issues. Figure 4-1 illustrates the reduced solubility of manganese and iron as pH increases. The soluble forms of iron and manganese are represented by Fe^{+2} and Mn^{+2} on the diagram. A phosphate-based sequestering agent is usually used in these situations or at lime softening plants to reduce calcium deposition on filter material and distribution systems.

Caustic soda, caustic potash, soda ash, potash, limestone contactors, and aeration (where sufficient dissolved CO_2 exists in the water) are the principal methods for increasing the pH. Soda ash, potash, and limestone contactors also increase DIC; aeration decreases DIC in the process of increasing pH.

Caustic soda/caustic potash. When caustic soda (sodium hydroxide) or caustic potash (potassium hydroxide) is used, dosing must be strictly controlled. Caustic is recommended only for waters with DIC greater than 5 mg/L. In low-alkalinity waters, even a small overdose of caustic will increase pH to very high values. For low-alkalinity waters, caustic is recommended only when used in conjunction with carbon dioxide. Caustic soda is usually delivered as a 25 or 50 percent solution. Caustic soda is a hazardous chemical, and severe burns can occur on exposure. It also has a low freezing temperature. A 50-percent solution has a freezing point of approximately 53°F (14°C), and a 25-percent solution has a freezing point of around 0°F (–18°C). Caustic feed systems, at a minimum, should include an eye washing system, full shower, eye goggles, protective gloves, boots, aprons, easy-to-handle barrels, and chemical containment areas.

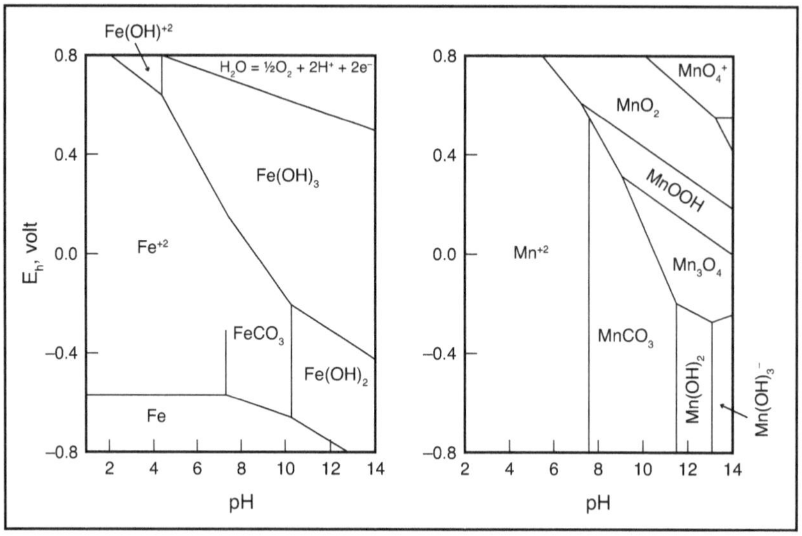

Figure 4-1 Iron and manganese solubility

Soda ash/potash. Soda ash (sodium carbonate) and potash (potassium carbonate) are dry compounds that are relatively safe to handle compared to caustic. These carbonate chemicals will not cause skin irritation. They are recommended for raw waters with DIC levels between 2 and 25 mg/L. When soda ash or potash is added to a water, there is an increase in DIC as well as pH. Additionally, a 1-mg/L dose of soda ash or potash increases the alkalinity by approximately 0.94 mg/L as $CaCO_3$. Because soda ash and potash are safe to handle, they are strongly recommended as the pH adjustment chemical for schools, condominiums, or any facility where technical resources are limited. They dissolve more easily than lime. Potash is more expensive than soda ash but is more soluble and easier to handle. Therefore, many very small water systems have found potash easier to implement than soda ash.

Sodium bicarbonate. Sodium bicarbonate is a nonhazardous chemical and is easy to handle. This chemical is applied to waters with DIC less than 5 mg/L. When sodium bicarbonate is added to a water, there is an increase in DIC as well as pH. Additionally, a 1-mg/L dose of sodium bicarbonate increases the alkalinity by approximately 0.59 mg/L as $CaCO_3$. This chemical is limited, however, in that it cannot increase pH above 8.3. Some utilities use both soda ash or caustic soda and sodium bicarbonate together if a significant increase in pH and alkalinity is needed.

Lime. Lime is commonly available as either hydrated lime ($Ca(OH)_2$) or quicklime (CaO). The use of quicklime requires a lime slaker to produce hydrated lime. Quicklime is usually used in larger facilities because the lower cost of chemicals offsets the higher capital and labor costs. Notably, lime can be difficult to use because of its low solubility and high hygroscopy (i.e., attraction to moisture in the atmosphere). Caution should be taken to minimize moisture in quicklime storage areas and silos. Most quicklime is used within a few weeks of chemical manufacture, as it may form calcium carbonate cake by absorbing moisture and carbon dioxide. Dry hydrated lime is less hygroscopic and is easier to store than quicklime. Specifications usually require quicklime and hydrated lime to contain 90 percent and 68 percent available calcium oxide, respectively.

Lime may contain contaminants such as heavy metals including aluminum. Depending on the quality of lime and the amount used, it is possible to increase the aluminum level by as much as 0.5 mg/L. Lime may also lead to excessive finished water turbidity if the lime does not dissolve completely. Some utilities will use lime for gross

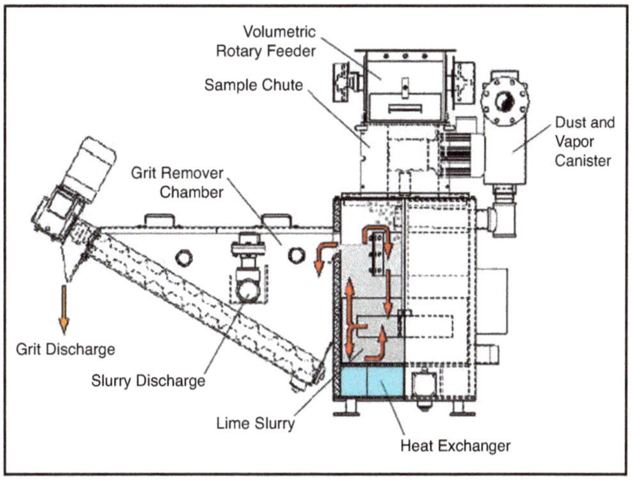

Figure 4-2 Slurry slaker

pH adjustment and then fine-tune the pH with caustic soda. In addition, lime often has a considerable amount of grit, and sand must be removed once the lime is fully dissolved.

Most water utilities will use a lime slaker. There are basically three types of lime slakers available on the market:

1. Slurry detention slakers
2. Paste slakers
3. Ball mill slakers

A slaker must mix the correct amount of quicklime (CaO) and water, must hydrate the quicklime, and must separate the impurities and grit from resultant calcium hydroxide slurry.

A slurry slaker generally uses an initial lime-to-water ratio ranging from 1 to 3.3 to a ratio of 1 to 5 depending on the equipment manufacturer and quality of CaO and water. Typically a slurry slaker, sometimes called a *detention-type slaker*, is comprised of two chambers. The first chamber is called the *slaking chamber*, where lime and water are mixed. The second chamber is usually used as a *grit removal chamber*. The lime slurry flows by gravity from the first chamber to the grit chamber. The slurry viscosity is reduced in the second chamber by the addition of cold water to allow the heavier grit to settle to the bottom of the second chamber where the grit is elevated and discharged by a screw. Figure 4-2 shows one such slaker.

The slurry slakers are generally designed for a retention time of 10 minutes at full rated capacity. This retention time means that from the time a particle of CaO enters the slaker until it exits into the grit remover takes an average of 10 minutes.

Slurry slakers are available in a variety of sizes ranging from 150 lb/hr to 15 tons/hr. Slurry slakers are also available with external vibratory grit separation screens. Because the technology for slurry or detention-type slakers has existed longer, those slakers are the most common types in Europe and the United States. Paste, and more recently batch, slakers are becoming more common.

Paste slakers, as the name implies, slake the lime in a paste form. The lime-to-water ratio is generally 1 to 2.5. Paste slakers are compact in size and are designed for a retention time of 5 minutes in the slaking chamber. In a paste slaker, because the hydroxide paste is too heavy to flow by gravity, a pair of horizontal rotating paddles pushes the paste forward toward the discharge point. Once the paste exits the slaking chamber, it is diluted to

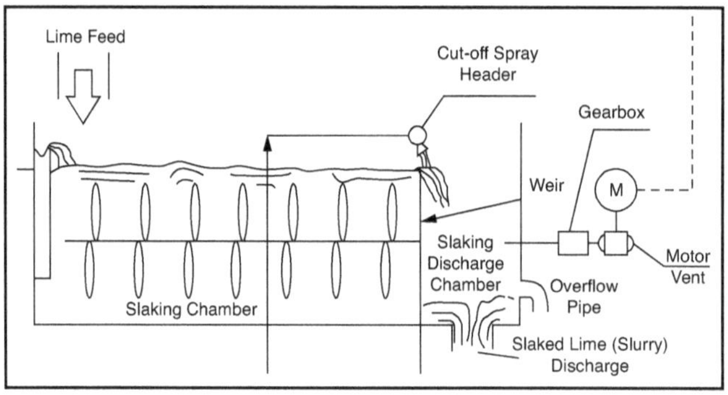

Figure 4-3 Paste slaker

approximately 1 part lime to 5 parts water. This dilution allows grit separation by gravity or by an external vibrating grit screen. The slurry consistencies from a paste slaker and a slurry slaker are the same after dilution for grit removal.

Paste slakers are available in sizes ranging from 1,000 lb/hr to 8,000 lb/hr. Paste slakers are mostly used in the United States. Figure 4-3 depicts a paste slaker.

Ball mill slakers are an adaptation of ball mills, which were originally designed for wet and dry grinding to the job of slaking. Two types of ball mills are used for slaking, horizontal and vertical. Ball mill slakers are generally used where

- capacity required is too large for other types of slakers;
- because of zero discharge conditions at the site, no grit discharge is allowed; and
- when water available is too high in sulfates or sulfites for regular slakers.

The ball mill slakers are much more expensive than paste or slurry slakers. They are available in sizes ranging from 1,000 lb/hr to 50 tons/hr. Figure 4-4 shows an attritor-type vertical ball mill lime slaker. The ball mill slakers are equipped with an external classifier, which separates slurry from the oversized grit and impurities. The oversize grit is recycled back into the mill for regrind.

Limestone contactor. A limestone contactor is usually an enclosed filter containing crushed high-purity limestone. As the water passes through the limestone, the limestone dissolves, raising the pH, calcium, alkalinity, and DIC of the water. The system is very simple and requires little maintenance and, as such, is typically used by small systems (Spencer 1998, Letterman et al. 1986). Several references are available for determining if limestone contactors are suitable for certain water qualities (Letterman and Kathari 1996; L. Benjamin et al. 1992; Letterman et al. 1991, 1986); however, in general, a water system should have pH less than 7.2, calcium less than 60 mg/L, and alkalinity less than 100 mg/L as $CaCO_3$ (USEPA 2016). Additionally, several other factors, such as temperature, iron, manganese, and aluminum, can impact limestone dissolution and need to be carefully considered when designing and operating a limestone contactor (Jalil 2003).

Aeration. Aeration is the only pH adjustment method that does not add a chemical to the water and the only one that can reduce excess DIC by removing oversaturated carbon dioxide. Stratified surface water sources as well as most groundwater sources are good candidates for aeration for corrosion control treatment. A stratified surface water has different density and water quality layers as a result of temperature changes over the summer season. The bottom layers often lose dissolved oxygen and have elevated levels of carbon dioxide, iron, and manganese because the stratification prevents diffusion of dissolved oxygen from the upper water to the lower. Many groundwater systems have low to moderate levels of alkalinity but low pH and high DIC values because of the presence

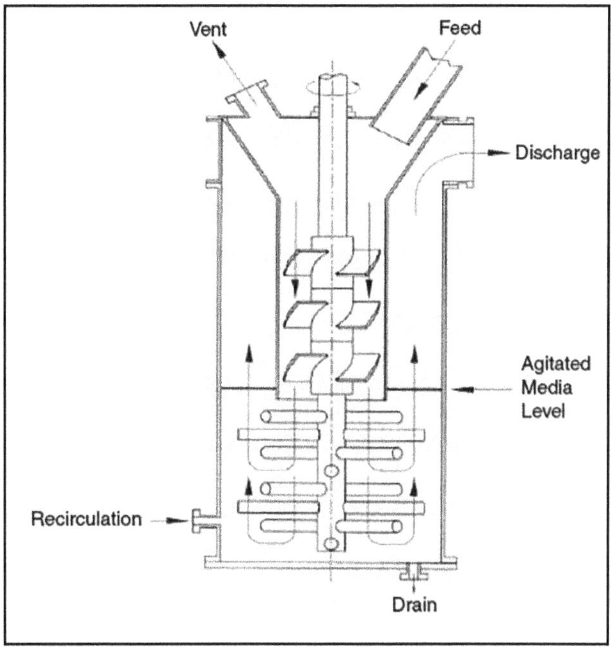

Figure 4-4 Ball mill lime slaker

of dissolved carbon dioxide at levels exceeding saturation values. The extent to which the saturation point is exceeded is critical to determining the increase in pH achievable by aeration (Lytle et al. 1998).

There are a wide variety of aeration systems, including diffused bubble systems, packed towers, tray systems, spray systems, and Venturi systems. Any aeration system selected for pH adjustment should be capable of removing at least 80 to 90 percent of the carbon dioxide. Larger amounts of pH adjustment will require the use of designs that produce higher percentages of carbon dioxide removal. pH values above 7.2 may be difficult to achieve by aeration, depending on the amount of DIC in the water (Lytle et al. 1998). If higher values are desired, chemical adjustment will be necessary. One of the disadvantages associated with aeration is that repumping of the water is required. Some water systems can configure their well, plant, and storage locations to maximize the use of gravity in the hydraulics of their distribution networks. Some state regulatory agencies require systems to disinfect the water after aeration so that any microbes introduced during aeration will not grow out in the distribution system. Aeration may also oxidize iron and manganese, resulting in red- or black-water problems in the distribution system if the levels of iron and manganese are high enough and filtration is not provided (Lytle et al. 1998).

Carbon dioxide. Carbon dioxide in water dissociates to carbonic acid (H_2CO_3), which will decrease pH when added. However, it also provides carbonate to the water, which will provide additional alkalinity when followed by a base chemical. For corrosion control purposes, carbon dioxide is typically added in extremely low-alkalinity waters prior to caustic. This will provide adequate alkalinity and buffer intensity to the water leaving the plant and minimize extreme elevated pH levels during a caustic overdose.

Corrosion Inhibitors

The use of corrosion control inhibitors is common in the drinking water industry. However, each chemical has different properties and may not truly inhibit corrosion. Some chemicals simply mask the effects of corrosion to prevent aesthetic problems, such as red or black water. Currently, the most common inhibitors used in the drinking water industry are orthophosphates, polyphosphates, blended phosphates, and silicates. The inhibitors can be classified as sequestering agents, anodic coaters, or cathodic inhibitors. Except for those chemicals that act as sequestering agents, the chemicals form a protective, relatively insoluble film on the inside surface of the pipe, acting as a barrier between the water and the pipe surface. Stannous, or tin, chloride has been tested for the control of lead corrosion but has not been widely used in full-scale applications.

The success of an inhibitor depends on three factors: dosage, feeding, and flow rates. The initial dosage of any inhibitor should be three to five times the normal dosage of inhibitor to build up a protective film as fast as possible (M.M. Benjamin et al. 1990). Building up the film may take weeks to months depending on the system. The inhibitor should be fed continuously. Interruptions in feed may cause a breakdown in the scale and prevent reformation of the protective barrier. The flow rates must be sufficient to continuously transport the inhibitor to all parts of the metal surface; otherwise, an effective protective film will not form. For example, inhibitors are not effective for controlling corrosion of tanks because the coating cannot be fed continuously to all parts of the tank surface.

The choice of inhibitor and the necessary dose are dependent on the water system, operational conditions, distribution water quality, and particular corrosion concerns. Phosphate-based inhibitors are, by far, the most commonly used chemical inhibitors (Figure 4-5). Two surveys, one in 1994 (Reiber et al. 1997, Dodrill and Edwards 1995) and one in 2001 (McNeill and Edwards 2002), examined the use of corrosion inhibitors by US drinking water utilities. Both surveys found that more than half of all utilities serving more than 10,000 people use some type of phosphate-based corrosion inhibitor. The earlier survey found that as the size of utilities decreased, so did the frequency of inhibitor use.

McNeill and Edwards (2002) also noted a shift from the use of polyphosphate to orthophosphate following implementation of the Lead and Copper Rule and noted the potential for polyphosphates to increase dissolved metals concentrations. The utilities surveyed also cited a variety of reasons for using phosphate-based inhibitors (Figure 4-6). The most commonly cited reason was lead and copper corrosion control, followed by prevention of red water. But other factors, such as protection of cement mortar linings and prevention of calcium scale buildup, were also noted. The doses reported in the survey ranged from less than 0.2 mg/L as P to greater than 3.0 mg/L.

Orthophosphate. Orthophosphate (PO_4^{-3}) added as a corrosion control treatment chemical can combine with lead and copper in plumbing materials to form a wide variety of insoluble compounds, which is an effective corrosion control strategy known as *passivation*. Chapter 2 discusses the mechanism of how inhibitors reduce corrosion. In laboratory studies, Edwards and McNeill (2002) found that orthophosphate dosing at 1 mg P/L was able to reduce soluble lead release by approximately 70 percent in most cases. Doses of orthophosphate applied in the United Kingdom are generally two to four times those commonly applied in the United States and are highly effective. These doses, in the long term, tend to resolve both particulate and dissolved lead release (Hayes et al. 2008, Cardew 2009). It has also been observed that, in the case of some new pipes, lead release increased. The authors theorized that this finding may have been because pH increase resulting from orthophosphate stagnation was less in new pipes. Reiber (2006) noted that orthophosphate was effective for hardening existing iron scales at pH 7.4–7.8 and can help to prevent red-water occurrences. Similar results were reported by Lytle et al. (2003), who observed that total iron released remained low following discontinuation of orthophosphate addition due to the formation of iron-phosphorus solids in the scales, thereby reducing the solubility of ferrous iron and/or decreasing the permeability of the scales.

Orthophosphate is available as straight phosphoric acid, and there are also neutral pH orthophosphate products, such as potassium orthophosphate. Orthophosphate is significantly less expensive than polyphosphates or zinc orthophosphate.

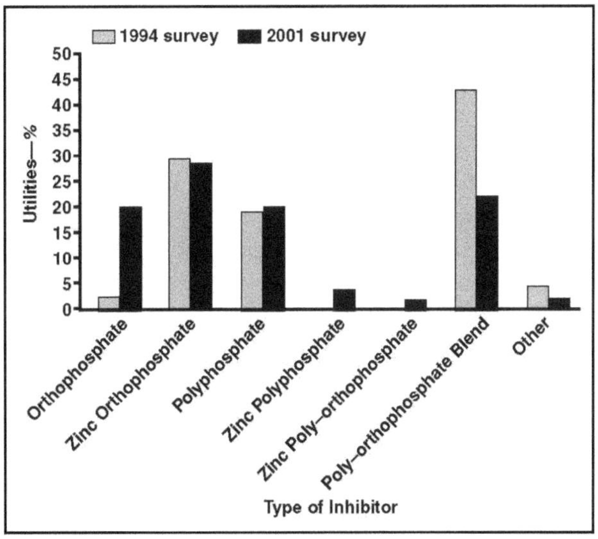

Source: *Rushing et al. 2002*

Figure 4-5 Type of phosphate inhibitor used by water utilities

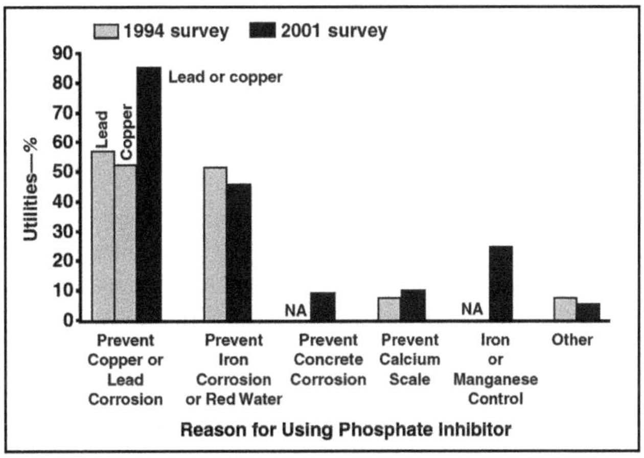

Note: Utilities were allowed to select more than one answer, so values do not total 100%. NA—not applicable; utilities in the 1994 survey were not asked about the reason for using inhibitors.

Source: *Rushing et al. 2002*

Figure 4-6 Reasons for using phosphate inhibitors

Orthophosphate works best in a pH range of 7.2 to 7.8 (Brown et al. 2013), although some studies have indicated that orthophosphate can still reduce lead in the range of 7.0 to 8.0. Pipe loop studies have demonstrated an immediate increase in lead concentrations whenever the pH of a water fell out of this range. Normal dosages are 0.5 to 3 mg/L as PO_4. Small doses of orthophosphate appear to significantly reduce iron release, but larger doses around 3 mg/L as PO_4 are required for iron corrosion control.

Zinc-containing phosphates. Zinc orthophosphates were first used for corrosion control in 1950. Zinc orthophosphates as a technology provide effective and comprehensive corrosion inhibition. Traditionally, they have been used to stop colored water caused by bleeding tuberculation and have a long history of success. This technology is very effective in aggressive waters with no source iron and manganese to sequester and where calcium stabilization and scaling are not concerns. Zinc orthophosphates have demonstrated effectiveness in providing lead and copper protection. Zinc orthophosphates also reduce asbestos fiber counts and protect cement pipe. Most formulations are proprietary and consist of a zinc salt (either chloride or sulfate) combined with orthophosphate, formulated in a wide variety of zinc-to-phosphate ratios ranging from 2:1 to 1:10. They function by two distinct mechanisms: film formation and electrochemical passivation. Zinc orthophosphate works best in waters with a pH between 7.3 and 7.8. Zinc orthophosphates should never be fed to a water with a pH above 8.1 because of premature zinc precipitation. In general, zinc orthophosphates are less soluble and dissolve less readily than polyphosphates or phosphoric acid. Zinc orthophosphate may perform better at lower dosages than orthophosphate, particularly in harder waters. The scales of the protected pipes appear to have equal parts zinc and phosphate, leading to the belief that the zinc and phosphate have a synergistic effect. It is postulated that the zinc forms carbonate compounds and protects the cathodic sites while the phosphate forms a protective barrier on the anodic sites. Several studies, however, have found no additional benefit as a result of the presence of zinc (McNeill and Edwards 2000; Williams 1990).

Zinc polyphosphates are also available. This technology is appropriate for harder, less aggressive waters that require some level of sequestering and/or calcium stabilization. Zinc polyphosphates offer a variety of benefits. In addition to some of the traditional benefits offered by zinc orthophosphates, the polyphosphate component prevents scale and stabilizes soluble iron, manganese, and calcium by sequestering these ions. Limited lead protection is obtained through hydrolysis of the polyphosphate to orthophosphate ion. Products in this category consist of a zinc salt combined with hexametaphosphate in a dry state. As with polyphosphates, zinc polyphosphates should not be used for lead corrosion.

Zinc orthophosphates and zinc polyphosphates contribute to the zinc loads at downstream wastewater treatment facilities and, therefore, should be considered when selecting a corrosion inhibitor. Drinking water systems that use zinc-based inhibitors and discharge to wastewater systems with zinc limits may be limited in the dose of inhibitor that can be applied, or wastewater treatment plants may require modifications and/or added operational costs to address zinc accumulation in the process biosolids. The zinc in zinc orthophosphates and zinc polyphosphates may improve corrosion control in harder waters; however, the zinc-containing products do not appear to offer additional corrosion control for most waters compared to other orthophosphate compounds.

Polyphosphates. Polyphosphates can sequester soluble metals and prevent or delay their precipitation, reducing colored-water episodes. Reiber (2006) noted that polyphosphate can be effective for the control of red water by mobilizing friable iron and helping to stabilize the corrosion scale core. In one utility, the use of polyphosphates was credited with reducing the number of red- and black-water complaints from as many as 35 per month to less than 5 per month in less than 1 month (Arweiler et al. 2003). Holm and Schock (1991) also noted that the complexation properties of polyphosphates can be effective for the control of iron (red water), manganese (black water), and carbonate scaling but may also significantly increase the solubility of lead in home plumbing systems. Other studies have also found polyphosphate to accelerate copper and lead corrosion in most waters. Dodrill and Edwards (1995) reported that the 90th percentile lead release when dosing polyphosphate at 1 mg P/L was more than double that when dosing orthophosphate in utilities with alkalinities less than 30 mg/L and pH less than 7.4. Further, under similar pH and alkalinity conditions, lead release using polyphosphates was greater than when dosing no inhibitor at all. In the same alkalinity range but at pH 7.4 to 7.8, polyphosphate actually performed better than orthophosphate, and both performed better than having added no inhibitor at all. At alkalinities 30 to 74 mg/L and pH 7.4 to 7.8, there did not appear to be any significant difference in orthophosphate and polyphosphate performance. In general, polyphosphates may be used as a sequestering agent for iron, manganese, and calcium but should not be used for lead control. A note of caution: Even if polyphosphates are used with the intention of sequestering, lead and copper may be held by the chemical and increase in the water.

Polyphosphates are available in multiple forms. Sodium hexametaphosphate, for instance, is effective for red-water control (McNeill and Edwards 2001), but sodium hexametaphosphate is not typically effective as a lead corrosion inhibitor. Edwards et al. (2001) found that with few exceptions, soluble lead concentrations were higher in systems dosed with hexametaphosphate than without inhibitor. Under controlled laboratory conditions, Edwards and McNeil (2002) observed that soluble lead concentration increased by approximately 1.6 mg/L for each 1-mg/L increase in sodium hexametaphosphate.

Polyphosphates will revert to orthophosphate over time in the distribution system. The half-lives of most simple polyphosphates are on the order of a few days (Clesceri and Lee 1965). In pipe rig experiments, Holm and Edwards (2003) confirmed these results and found that 20 to 95 percent reversion occurred in as few as three days. For this reason, the results of many polyphosphate studies may be confounded by the presence of orthophosphate (McNeill and Edwards 2001).

Polyphosphates are cathodic corrosion inhibitors. Divalent cations such as calcium (Ca^{+2}) at levels of at least 10 mg/L are needed along with the polyphosphates for the polyphosphates to be effective. Positively charged colloidal complexes are formed that migrate to the cathode, forming an amorphous polymeric film. This film is self-limiting. Analysis of the film has shown ferric pyrophosphate and iron/calcium metaphosphate to be the primary constituents. The pH is best maintained at the level of 6.5 to 7.0 because as the pH drops, the rate of reversion of the polyphosphate to orthophosphate increases. Typical dosages are 2 to 4 mg/L as PO_4. In these applications, the polyphosphate is acting both as a reservoir of potential orthophosphate and as a cathodic corrosion inhibitor. More soluble and less likely to be precipitated in its polymeric form, polyphosphates can be retained longer in the system than orthophosphates.

Polyphosphate feed points should be separated from the chlorine injection point by as much distance as possible. The polyphosphate feed point should also be ahead of the chlorine injection point. If polyphosphate is fed after the chlorine, there is a possibility that the iron and manganese will be oxidized by the chlorine before the sequestering action can take place, causing iron and manganese precipitates to be pumped out into the distribution system.

The feed equipment used for phosphate addition is similar to the equipment used to feed fluoride, consisting of a storage tank, solution tank, feed pump, and controller to pace the equipment. The storage tank and the solution tank must contain at least 10 mg/L of free chlorine residual to prevent bacterial growth in the phosphate solution; polyphosphate is an excellent food source for bacteria.

Solutions of polyphosphates may also be made up from powder in a saturator similar to the one that is used in making up solutions of dry-fluoride compounds. Phosphate solutions containing more than 0.5 lb phosphate/gal (60 ppm) may be very viscous. It is important that any solution be fed within 48 hours of its production.

Blended phosphates. Numerous proprietary polyphosphate–orthophosphate blends are available and, in general, have shown good success for corrosion control because they can theoretically provide both anodic and cathodic corrosion inhibition; ensure the orthophosphate ion availability over a longer period of time; and control calcium, iron, manganese, and lead deposition. In one study, polyphosphate–orthophosphate blends reduced copper corrosion at rates greater than the rates when orthophosphate is used alone (Edwards et al. 2001). However, a study in Wisconsin by Cantor et al. (2000) looked at the effect of the use of blended phosphates on lead and copper pipe corrosion. Their results showed that using a blended phosphate (with polyphosphate contributing two-thirds of the phosphorous concentration and orthophosphate contributing the other third) increased lead concentrations to four times what they were in the untreated pipe loop.

Orthophosphate–polyphosphate blends are being produced specifically to use in water systems when an orthophosphate inhibitor is a viable corrosion-control approach and a polyphosphate is also required to meet other treatment objectives, such as the control of red-water discoloration from iron or hardness stabilization. Blended phosphates have the potential to provide corrosion control, finished-water stabilization, and distribution system protection. Testing for both orthophosphates and polyphosphates in the distribution system will assist in determining the correct inhibitor dosage.

According to USEPA's *Optimal Corrosion Control Treatment Evaluation Technical Recommendations for Primacy Agencies and Public Water Systems* (USEPA 2016):

> The effectiveness of blended phosphates [for lead and copper corrosion] cannot be based on the orthophosphate concentration in the blend for the following reasons:
>
> - Blended phosphates control corrosion by creating a barrier film from the interaction of calcium and aluminum in the bulk water with phosphorus containing compounds (Wasserstrom et al. 2015). Thus, calcium and aluminum play a role in effectiveness.
> - If the polyphosphate portion of the blend has a high affinity for sequestering lead or copper, it may counteract the benefit of the orthophosphate portion in forming solid lead and copper compounds.
>
> The percent of orthophosphate in the blend can vary widely (from 5 to 70 percent (Hill and Cantor 2011)). Blended phosphate should contain a minimum orthophosphate concentration of 0.5 mg/L as P (1.5 mg/L as PO_4) as a starting point for evaluation. The orthophosphate ratio in the blend and/or the dose may need to be increased to provide adequate lead control. In some cases, however, simply adding more blended phosphate may not be effective because, if there is excess polyphosphate available beyond what is bound up with other constituents in the water, it can sequester the lead and copper. EPA recommends a demonstration study, additional monitoring, or both for systems that recommend blended phosphates to control lead release.

Silicates. Sodium silicate (Na_2SiO_3) is used primarily as a sequestering agent for the control of iron and manganese but has been successfully used to control lead and copper corrosion. The primary corrosion protection offered by silicates appears to be the resulting increase in pH and formation of a protective film (Vik et al. 1996). For this reason, silicates are most likely to be used in soft waters of low pH and high oxygen concentrations, although they have been successfully applied in varying water qualities. With respect to lead and copper, sodium silicates are basic and may result in an increase in pH, which is usually beneficial from a lead and copper control perspective. It has also been suggested that silicates form a thin protective layer that acts as a diffusion barrier to lead and copper corrosion (Stericker 1945). The films are self-limiting and do not build up in thicker layers with increased dosages. The films also break down relatively quickly if there is an interruption in feed.

Sodium silicate is available either in a dry chemical form or as a liquid solution. Dosages range from 4 to 40 mg/L, depending on the application. For the passivation of lead and copper, the required dose of sodium silicate solution is typically much higher (20–30 mg/L as sodium silicate) than that of the phosphate-based inhibitors (1–2 mg/L as P).

The effectiveness of silicate inhibitors may be limited by the presence of multivalent cations such as calcium and magnesium (Robinson et al. 1987). Higher dosages are used in waters with higher concentrations of hardness, chloride, and dissolved solids. High-silicate doses may also be necessary in low-pH water, and as such, it may be more economical to raise the pH using caustic soda or soda ash prior to silicate addition (Vik et al. 1996). In general, it is recommended that an initial start-up dose of 24 mg/L be applied. The dose may be incrementally decreased after 60 days to maintain an 8-to-12-mg/L residual. It should be added at or shortly after oxidant addition.

In general, silicates are less effective for corrosion control than the use of zinc orthophosphate or orthophosphate alone. However, significant reductions in 90th-percentile lead and copper concentrations have been reported following the addition of silicates (Schock et al. 1998, 2005). In one study, 90th percentile lead concentrations were reduced from 0.077 mg/L to 0.002 mg/L, and 90th percentile copper concentrations were reduced from 5.87 mg/L to 0.27 mg/L following a change from a polyphosphate inhibitor to sodium silicate. This dramatic decrease may also have been contributed by the removal of the polyphosphate. In addition, silicates are very expensive and cost more than phosphates. As such, the high costs associated with silicates and the time required to reduce corrosion in the distribution system have eliminated the use of silicates as a possible corrosion treatment for many water treatment plants.

Phosphate–silicate blends. The use of phosphate–silicate blends has increased in the past few years. The blends appear to be as good as phosphate products alone but are used at reduced concentrations. Phosphate–silicate blends have also reduced corrosion in steel pipe.

Stannous chloride. Stannous chloride (also known as tin chloride) has been studied for its potential use for the control of lead and copper corrosion. The stannous ion is the in-between oxidation state of tin and may be either oxidized or reduced depending on the liquid and the ionic state of the materials nearby. In theory, stannous ions in the water are drawn to active metal surfaces that are constantly losing electrons as they oxidize or as they are forced into solution by electric potentials. The positively charged tin ion is drawn to these negative surfaces and, depending on conditions, reduces the oxidized ions (rust and corrosion) and also reduces itself. This action leaves behind an inactive layer of tin and a reducing aqueous layer of stannous chloride very close to the surface of the pipe. When an oxidizing agent attacks the tin surface, the stannous ions protect it, and if the tin surface is diminished, it is replaced by the ongoing water treatment. Corrosion is still occurring, but the corrosion is now to the tin layer and not to the pipe material.

Several bench- and full-scale demonstration studies have failed to demonstrate the use of stannous chloride as a viable corrosion control treatment method. One study investigated the use of stannous chloride for control of lead more thoroughly and found insignificant reduction in lead levels compared to other conventional treatments (Hozalski et al. 2010). A full-scale demonstration test conducted over a several-year period in Saint Paul, Minn., did show promise for the control of lead corrosion (Rezania 2015). However, it has been speculated that the toxicity of tin inhibited biofilm growth in the distribution system, and, as a result, the pH became more stabilized, contributing to lower lead concentrations. Additionally, a study in Brainerd, Minn., indicated that stannous chloride was not effective at controlling copper corrosion in a groundwater system with high DIC and high hardness (Rezania 2015). At this time, no water systems actively using stannous chloride for the control of lead or copper corrosion have been identified. Overall, additional data and research are needed before stannous chloride is approved for the control of lead or copper corrosion in drinking water systems.

CONSIDERATIONS FOR CORROSION CONTROL TREATMENT SELECTION

System Size

System size can impact the selected corrosion control treatment strategy. Systems with limited resources (i.e., very small systems) may not have the necessary funds and/or operational/maintenance capabilities to implement a fully optimized strategy, but they may still be able to implement some changes that can reduce corrosion sufficiently. For example, a small system may elect to adjust pH only and not adjust alkalinity to reduce implementation and chemical costs. This partially optimized strategy may be sufficient to comply with lead and copper regulations and, as such, can be considered a suitable corrosion control treatment strategy, given system constraints.

Control of Microbiologically Influenced Corrosion

Microbiologically influenced corrosion (MIC) is discussed in chapter 2. To prevent and control MIC, it is important to understand the operating conditions of the system. Hence, evaluation of the biological and abiotic parameters is essential when implementing prevention and control measures.

Methods implemented to prevent MIC should address the following fundamental issues:

- Inhibition of the growth and/or metabolic activity of microorganisms
- Modification of the environment in which the corrosion process takes place to avoid adaptation of microorganisms to the existing conditions

Techniques for controlling MIC include disinfection, control of water age, flushing, and nutrient control, which are discussed further in chapter 5.

Secondary Disinfectants

Chlorine. Free chlorine residuals can impact the metal oxidation rates and the nature of scales that form on the interior of distribution system piping. Chlorine is frequently used in a gaseous form, which hydrolyzes to form hypochlorous acid (HOCl) and hydrochloric acid (HCl):

$$Cl_2 + H_2O \rightarrow HOCl + HCl$$

Hypochlorous acid is the active species, and its dissociation is pH dependent:

$$HOCl \rightarrow H^+ + OCl^-$$

At pH 7.5, the concentrations of hypochlorous acid and its ions are equal. At more alkaline pH values, the equilibrium is displaced in favor of the ionic form. At pH 9.5, all chlorine is converted to hypochlorite ions, which have poor biocidal activity. Based on this pH-sensitive equilibrium, the range from pH 6.5 to 7.5 is considered to be ideal for the action of chlorine for use as a biocide and to control MIC (Videla 2002). Chlorine at lower pH levels can accelerate corrosion.

Chlorine is an excellent algaecide and bactericidal agent, despite the evidence that the effective concentration of chlorine is considerably reduced following the penetration of bacterial biofilms. Measurements performed using selective microelectrodes showed that the chlorine concentration inside the biofilm represented only 20 percent of the concentration measured in the bulk liquid in contact with the biofilm. This finding may explain the decrease in chlorine biocidal efficacy when it is used on biofilms developed by different species of sulfate-reducing bacteria or by mixed bacterial consortia (Videla 2002).

Other sources of chlorine are salts of hypochlorous acids, such as sodium hypochlorite (NaOCl) or calcium hypochlorite ($Ca(OCl)_2$). Their action is similar to that of gaseous chlorine for disinfection, though these chemicals are mildly basic and might raise the water pH away from the pH 6.5–7.5 range for optimized MIC control. Chlorine dioxide (ClO_2) is also used as a substitute for gaseous chlorine for primary disinfection, typically in smaller systems, because it produces fewer chlorinated, regulated disinfection byproducts (DBPs) than chlorine. However, it does produce both chlorite, which is regulated in Canada and the United States, and chlorate, which is regulated in Canada (Health Canada 2008) and may be regulated in the near future in the United States. This chemical provides the same level of disinfection without any pH impacts.

The effects of free chlorine on the corrosion of copper in drinking water have been studied more thoroughly than the effects of other disinfectants because chlorine is a more common disinfectant used in drinking water distribution systems. Boulay and Edwards (2001) summarized the relevant results from the 1980s and found that copper corrosion may be increased or decreased by chlorine depending on pH. At low pH, copper leaching caused by chlorine was found to be the highest.

Recent research and investigation of the events in Washington, D.C., have observed that high free chlorine residuals inhibit lead corrosion as the resulting elevated oxidation–reduction potential (ORP) contributed to the formation of an insoluble Pb(IV) scale on the pipe wall, inhibiting lead release (Lytle and Schock 2005). Under this mechanism, a high free chlorine residual must be maintained throughout the distribution system or the insoluble Pb(IV) scale will be reduced to the more soluble Pb(II), resulting in higher dissolved lead concentrations. Maintaining an adequate residual may be difficult in systems with long water age and may require increased chlorine addition, which can increase DBPs or result in potential taste and odor issues. Studies have also found that the presence of orthophosphate can inhibit the formation of new Pb(IV) scales, thus limiting the potential effectiveness of this strategy (Brown et al. 2013, Lytle and Schock 2005). As such, the use of high free chlorine residuals by itself is not recommended as the primary corrosion control technique, but it could be a good supplement to a primary corrosion control strategy.

Chlorinated systems with high THM levels may not be able to increase pH sufficiently for corrosion control treatment as a result of the increased formation potential for THMs and, therefore, may elect to implement a phosphate-based inhibitor. Alternatively, systems with high haloacetic acid (HAA) levels may not be able to decrease pH if needed to achieve the optimal range for orthophosphate addition (typically 7.2–7.8) because of the increased formation potential for HAAs.

Chloramines. Many utilities have changed from free chlorine to chloramines for secondary disinfection to comply with the Disinfectants and Disinfection Byproduct Rules. Chloramine is formed when ammonia is bound to chlorine. The chlorine-to-ammonia ratio ranges from 3:1 to 5:1. Because chloramines are not as strong an oxidant as chlorine, the typical chloramine residuals range from 2 to 4 mg/L. Chloramines are also more effective at the higher pH levels.

Changes from free chlorine to chloramines for secondary disinfection may, under certain conditions, also result in an increase in lead and/or copper solubility. The problem is not that chloramines inherently take up more metals; rather, the problem is the loss of the protective nature of free chlorine. Chlorine provides a higher ORP while chloramines have a lower ORP. Reiber (2004) reported that a change in disinfectant could impact the redox potential of the service line surface. ORP determines the primary mineral species of metal that will be present. Changes in ORP that cause a shift in mineral species from less soluble to a more soluble species, from Pb^{4+} to Pb^{2+} for example, may increase lead solubility. However, a survey by Awwa Research Foundation (Dyksen et al. 2007) has indicated that most utilities that switched to using chloramine as a secondary disinfectant did not experience an increase in corrosion.

Utilities using chloramines must be more vigilant with distribution system maintenance and monitoring. Ammonia is a nutrient that may be utilized by ammonia-oxidizing bacteria resulting in nitrification. In addition to the potential health effects of high nitrite levels and reduced disinfectant residuals, nitrification can lead to increased corrosion through microbial activity and lower pH. From 1999 to 2003, the City of Ottawa, Ontario, Canada, evaluated several chemical alternatives to control corrosion (Douglas et al. 2004). A corrosion control strategy was established and included increasing pH to 9.2 with a minimum alkalinity target of 35 mg/L as $CaCO_3$ (achieved through the use of sodium hydroxide and carbon dioxide). During the initial implementation phase, a pH of 8.5 was maintained, but customer testing followed by a wider investigation found an area of the city with high levels of lead at the tap (10–15 μg/L for flowing samples). The problem was attributed to nitrification within the distribution system, which caused a reduction in pH from 8.5 to a range of 7.8 to 8.2 and resulted in lead leaching from lead service lines. The pH was increased from 8.5 to 9.2 and this increase reduced lead concentrations at the tap to a range of 6 to 8 μg/L for flowing samples.

Reoptimization of Corrosion Control Treatment

Utilities are encouraged to reevaluate corrosion control treatment whenever changes are encountered in the source water or when utilities are contemplating treatment modifications such as changes to coagulant type or secondary disinfectant. Table 4-4 lists the impact on corrosion by changes to (1) source of supply, (2) water treatment, and (3) distribution system activities.

Utilities change treatment processes for a variety of reasons, including

- changes in source water quality;
- new or additional source water;
- improved cost-effectiveness of existing treatment;
- new or more stringent regulatory requirements;
- the need to meet or exceed customer expectations (e.g., industrial water user requirements);
- aesthetic parameters, such as hardness, color, and taste and odor;
- system pressure; and
- direction and flow of water.

The change of a process may have a significant effect on finished water quality parameters that impact corrosion. Treatment processes that impact the treated water pH and/or alkalinity are most likely to cause changes in corrosion chemistry. Lime softening, membrane processes, and ion exchange are three processes likely to impact these parameters. Other processes that may have negative corrosion impacts include ozonation, granular activated carbon, and biofiltration (AWWA 2005).

Changes in distribution system operation practices may also impact water quality and the effectiveness of corrosion control treatment. The operational issues most significant to maintaining water quality and effective corrosion control are

- water age,
- blending finished waters of different qualities,

Table 4-4 Potential impact of treatment changes on corrosion control

	Changes in Corrosion-Related Water Quality Parameters					Potential Impacts on Existing Scale			Other Corrosion-Related Impacts					
	Change in finished water pH	Change in finished water alkalinity	Change in Cl:SO_4	Change in finished water NOM	Change in finished water TDS	Change in DO or other dissolved gas concentration	Change in stability/solubility	Change in ORP/oxidation state	Physical disruption	Interference w/ corrosion inhibitor effectiveness	Galvanic corrosion	Pinhole leaks	Increased potential for MIC	Introduction of particulate lead into system piping
Source of Supply Changes														
Change in source water quality	•	•	•	•	•		•			•		•		•
Addition of a new source of supply	•	•	•	•	•		•			•		•		•
Blending of different source waters	•	•	•	•	•		•			•		•		•
Treatment Changes														
Change in free chlorine dose	•	•					•	•				•		
Change from chlorine gas to hypochlorite	•	•					•	•		•				
Addition of other oxidants/ disinfectants (O_3, $KMnO_4$, ClO_2)			•	•		•								
Conversion from free chlorine to chloramines	•						•	•		•				
Enhanced coagulation for NOM removal		•	•	•			•			•				
Change in coagulant type or dose	•	•	•	•			•	•		•				
Change in finished water pH	•	•					•	•		•		•		
Change in finished water alkalinity	•	•					•	•		•		•		
Addition of a corrosion inhibitor							•					•	•	
Change in inhibitor type or dose							•					•	•	
Addition/discontinuation of softening	•	•	•	•	•		•	•		•		•		
Addition of NF/RO	•	•	•	•	•		•	•		•		•		
Addition of GAC				•								•	•	
Use of biofiltration				•		•						•	•	
Distribution Operations and Maintenance Activities														
Lead service line replacement									•		•			•
Meter and other device replacement									•		•			•
Blending of different finished waters	•	•	•	•	•		•	•		•				
Flushing									•					•
Storage tank/reservoir maintenance									•					•

DO = dissolved oxygen
GAC = granular activated carbon
MIC = microbiologically influenced corrosion
NF/RO = nanofiltration/reverse osmosis
NOM = natural organic matter
ORP = oxidation–reduction potential
TDS = total dissolved solids

Adapted from AWWA 2005.

- system pressure, and
- direction and flow of water.

Water quality deterioration in a distribution system is often closely associated with water age within the distribution system. Changes in distribution system operations that result in significantly higher water age may have negative impacts on corrosion-related parameters, such as pH (Kirmeyer et al. 2000).

Secondary Effects of Corrosion Control Treatment

Implementation or reoptimization of corrosion control may have secondary impacts. These potential impacts must be considered when selecting a corrosion control method. Table 4-5 lists the potential of secondary impacts of corrosion control.

To minimize potential secondary impacts of corrosion control, it is best if incremental changes in finished water quality are made during startup to avoid exposing the system to large finished water quality changes over a relatively brief time. Rapid changes in water chemistry may disrupt the system equilibrium and affect the films and scales on the pipe walls (Kirmeyer et al. 2000). For systems adjusting pH/alkalinity, a rapid change would constitute making a 0.1 to 0.2 adjustment in pH units per week in systems with a high potential for secondary impacts to occur and 0.5 pH units per week in systems with low potential for secondary impacts (Leitch 1996). For phosphate addition, dosages should be made incrementally, approximately 0.2 to 0.5 mg/L as PO_4 over at least several weeks. If pH needs to be reduced for orthophosphate usage, the decrease should also occur incrementally over several weeks. To minimize the risk of adverse impacts, the utility should also avoid making other treatment changes during corrosion control implementation.

In general, secondary impacts of corrosion control treatment may include increased red water, increased post-precipitation of solids, reduced disinfection efficiency, changes in DBP formation, increased microbial growth, increased loadings on wastewater treatment plants, and additional operator requirements.

Red water. Red water, also known as yellow or colored water, results from the corrosion of iron-containing pipes used to transport water. Red water may be a secondary effect when initiating corrosion control. Changes to pH and alkalinity can increase iron corrosion through the softening of the iron tubercles. The introduction of polyphosphates or orthophosphates may lead to the release of corrosion by-products. Systems that experience red-water complaints following installation of corrosion control usually have a concentrated area or large numbers of unlined cast-iron mains. To minimize red-water issues during the initial implementation phase of corrosion control treatment, it is recommended that an aggressive flushing program be conducted prior to corrosion control treatment implementation to remove sediment, loose corrosion by-products, and biofilm. Once corrosion control treatment is implemented, if the alkalinity is greater than 40 mg/L and stable, DIC greater than 15 mg/L, and the pH is maintained between 7.5 and 9.5, red-water issues should be minimal and may be resolved with flushing.

Post-precipitation of solids. Increased pH and/or alkalinity has the potential to result in additional calcium carbonate precipitation. The addition of phosphate also has the potential to result in precipitation of phosphate complexes, such as aluminum phosphate, where metals are present in the finished water supply. It can also cause milky or cloudy discolored water, particularly when adding orthophosphate at doses above 3 mg/L as PO_4 (Tesfai et al. 2006). Additional scaling may lead to loss of hydraulic capacity and, in certain cases, decreased water pressures, reduced fire flow, and/or increased pumping costs. Increased scaling may also require additional maintenance and/or replacement of valves and other equipment in the distribution system.

Table 4-5 Potential secondary impacts of corrosion control treatment

Corrosion Strategy	Metals Release	Post-Precipitation of Solids	Disinfection	Trihalomethane	Haloacetic Acid	Microbiological Impacts	Turbidity	Color	Taste and Odor	Wastewater Treatment Plants
pH increase	Decrease in Pb, Cu, Zn. Possible short-term increase in alkalinity could lower iron uptake by increasing buffering capacity. Iron levels could be increased if pH is adjusted to the point of minimum buffer intensity.	Precipitation of iron solids, manganese, and calcium carbonate can occur at elevated pH.	Potential decrease in disinfection performance	Potential increase	Potential decrease	Post-precipitation of solids may create surfaces for microbial growth.	May increase because of post-precipitation of solids.	May increase because of post-precipitation of solids that may contribute to color.	Change in pH may affect formation of compounds resulting in taste and odor.	No change
Orthophosphate addition	Decrease in Pb, Cu, Fe, Zn. Adverse effect to cement compounds if pH decreases.	Precipitation of aluminum phosphate can occur if aluminum-based coagulant is used. Precipitation of calcium phosphate can occur at elevated pH. Precipitation of zinc carbonate can occur at elevated pH with the use of zinc orthophosphate.	No change if pH remains constant.	No change if pH remains constant.	No change if pH remains constant.	Zinc in some formulations may inhibit coliform growth. Possibly improved disinfection efficiency because of reduction in tubercles that provide a growth surface for bacteria. Potential increase in bacteria if phosphate is limiting factor.	Reduction in tubercles and increased disinfection efficiency could reduce turbidity. Post-precipitation of solids may contribute to turbidity.	Possible increase because of post-precipitation of solids that may contribute to color.	Potential growth of algae in open reservoirs may affect taste and odor.	Potential increase in phosphorous loading. Potential increase in zinc loading if zinc orthophosphate is used.
Polyphosphate addition	Possible increase in Pb, Cu, Fe, Zn, Ca	May reduce precipitation of iron solids by acting as a chelating agent. Calcium increase for asbestos–cement or cement-lined pipe and softening of cementitious lining.	No change if pH remains constant.	No change if pH remains constant.	No change if pH remains constant.	Unknown	Turbidity may develop at elevated pH with reversion of polyphosphate to orthophosphate (releases sequestered metals). Turbidity may be reduced by polyphosphate sequestering of iron and manganese.	Color may be reduced by polyphosphates sequestering iron and manganese. Reversion of polyphosphate to orthophosphate may cause post-precipitation of solids that contribute to color.	Potential growth of algae in open reservoirs may affect taste and odor.	Potential increase in phosphorous loading.
Silicate addition	Possible decrease in Pb, Cu, Fe, Zn, Ca	May prevent iron and manganese precipitation. Zinc carbonate may precipitate at elevated pH.	Potential decrease in disinfection performance due to pH increase.	Potential increase due to pH increase.	Potential decrease due to pH increase.	Potential reduction in overall corrosion may improve disinfection efficiency because of reduced tubercle formation. Post-precipitation of solids may provide growth environment for bacteria.	Turbidity may be reduced by silicate sequestering of iron and manganese.	Color may be reduced by silicates sequestering iron and manganese.	Elevated pH from use of silicates may cause formation of compounds affecting taste and odor.	No change

Source: *Kirmeyer et al. 2000*

Microbial growth. Increases in total coliform and heterotrophic plate counts may occur following implementation of corrosion control. A variety of bacterial types in relatively high numbers is found to be associated with tubercles taken from distribution pipe walls. Corrosion control may exert a physical or chemical attack on the tubercles, causing them to break from the pipe wall and releasing the bacteria into the bulk water. Flushing during implementation of corrosion control will aid in removal of the tubercles and sloughed biofilm, reducing the duration of increased bacterial levels in the bulk water.

There are concerns that the addition of phosphate-based corrosion control chemicals may cause an increase in bacterial levels because phosphate is a nutrient used by bacteria. Nevertheless, the use of phosphate-based corrosion inhibitors is associated with lower coliform levels (LeChevallier et al. 1996). Lowther and Moser (1984) reported that the levels of coliforms decreased within a few weeks following the application of zinc orthophosphate in the Seymour, Ind., system.

Disinfection efficiency. For disinfection, free chlorine is most effective at pH 6, and efficacy decreases at higher pH levels, suggesting that disinfection efficacy could be reduced if the pH is increased for corrosion control. Where possible, increases in the finished water pH for corrosion control should be implemented after primary disinfection is achieved so that the required dose of disinfectant to achieve the target pathogen inactivation credits would not be affected by corrosion control modifications. However, for those systems that use free chlorine for secondary disinfection, raising the pH for corrosion control could decrease secondary disinfection efficacy in the distribution system and necessitate a higher applied dose to maintain the target chlorine residual. Loss of secondary disinfection efficacy may result in increased microbial activity and lead to further loss of residual, taste and odor, and/or red-water complaints. Furthermore, loss of secondary disinfection efficacy could potentially reduce the likelihood of inactivation following a cross-contamination event in the distribution system or possibly exacerbate compliance challenges associated with total coliform requirements in the distribution system. Systems may need to slightly increase the secondary disinfectant dose to maintain the same disinfection effectiveness. This, however, may impact distribution system DBP concentrations, potentially jeopardizing compliance with THM and HAA regulations.

For those systems that use chloramines for secondary disinfection, monochloramine is the most effective disinfectant of the chloramine species and is the predominant species at pH 7 or greater. Above pH 7, the efficacy of monochloramine is relatively unchanged. As such, little impact to disinfection efficacy is anticipated as a result of pH modification for corrosion control.

Disinfection by-product formation. Corrosion control treatment strategies involving an increased finished water pH have the potential to result in higher THM concentrations. Conversely, strategies that require a lower pH, such as adjusting pH to optimize orthophosphate effectiveness, may result in higher HAA concentrations. Systems with elevated concentrations of THMs or HAAs may consider performing simulated distribution system testing to assess potential compliance issues with DBP regulations prior to implementing any changes to the finished water pH. It is also worth noting that select systems operating under approved filtration avoidance criteria may need to add filtration if changes in corrosion control treatment result in DBP compliance issues.

Wastewater treatment. When selecting a corrosion control treatment, it is important to investigate potential issues for local wastewater treatment plants. Many wastewater treatment plants have limits on the amount of phosphorus they may discharge. If a drinking water plant is considering the use of a phosphate-based corrosion inhibitor, it should investigate if the local wastewater treatment plants have the capability to treat the additional phosphorous loads. In certain cases, wastewater plants with existing biological treatment may need to add chemical treatment, and wastewater plants with existing chemical treatment may need to add a greater coagulant dose to comply with permitted phosphorous limits. The use of additional coagulant for chemical phosphorous removal would also result in greater sludge production and associated disposal costs. In areas of the country with very low nutrient limits, advanced treatment, such as membranes, may be necessary.

Zinc is also a concern for many wastewater treatment plants because most plants have limits on the amount of zinc that can be discharged into receiving waters. It is possible that the use of zinc orthophosphate may cause excessive zinc loadings, which the wastewater treatment plant may not be able to treat. The use of zinc-containing corrosion control chemicals may impact the biological processes at wastewater treatment plants, reducing the effectiveness of wastewater treatment.

Other secondary impacts. Many existing public water systems are very small groundwater systems (serving fewer than 3,300 people) that provide no treatment beyond chlorination. The addition of corrosion control treatment and associated chemical feed systems may require that these systems meet additional operator training, certification, and/or staffing requirements. For instance, in Ohio, community water systems serving fewer than 250 persons would be required to increase staffing from two days and one hour per week to three days and one-and-one-half hours per week.

BENCH TESTING

Bench testing can provide a relatively simple process to determine specific side effects of initiating a new corrosion control program, initiating other treatment changes, or checking the continuation of desired water system characteristics.

Regarding lead and copper control, three specific bench-scale tests are of interest for both surface water and groundwater systems:

1. **Calcium carbonate precipitation test.** This test should be performed before adjusting either pH, alkalinity, or calcium hardness to determine if excessive calcium scaling will occur because of the change. Excessive scaling causes hydraulic problems in pipes.
2. **ORP test.** This test should be performed before adjusting the concentration of disinfectant or changing the type of disinfectant. This is to determine if ORP will increase or decrease under the new conditions. An increase of ORP can aid in the formation of a protective plattnerite (lead oxide) scale and also creates an environment that is difficult for the growth of microorganisms.
3. **Phosphate test.** Phosphate added to water for lead and copper control can react with other chemicals in the water and form undesired precipitates. The precipitates use up the phosphate available for corrosion control and also can cause excessive scaling with hydraulic issues.

Calcium Carbonate Precipitation Test Procedure

A calcium carbonate precipitation test will provide an understanding of how much calcium carbonate could precipitate in a distribution system. This test is designed to help utilities avoid excessive precipitation and should be conducted if computer models during the desktop study predict a positive or near-positive calcium carbonate precipitation potential (CCPP) at the newly targeted pH, alkalinity, or hardness. This type of test will measure the true CCPP under real water quality conditions.

The following steps can be used to conduct a CCPP simulation:

1. Collect water from a source that is to undergo pH or alkalinity changes. This water is referred to as *baseline water.*
2. In a 1-L beaker, raise the pH and/or alkalinity of the baseline water to the required level using the desired pH adjustment chemical. Once achieved, continue slowly stirring the water in the beaker.
3. After 24 hours of mixing, remove a sufficient volume of water from the jar to analyze for either total calcium or total calcium hardness.
4. Remove another 50 to 100 mL of jar test water and filter through a 0.45-μm filter.
5. Measure the calcium or calcium hardness of the filtered water, which is the dissolved calcium in the water.
6. To obtain the amount of calcium that has precipitated, subtract the filtered or dissolved calcium from the unfiltered total calcium level.

Standard Method 2330 for calcium carbonate saturation can also be used to test whether a water has a tendency to precipitate or dissolve $CaCO_3$ (APHA et al. 2012). It should also be noted that polyphosphate, including sodium hexametaphosphate, can interfere with this test, so the test should be done before phosphate is added. Additionally, several spreadsheets or models, such as AWWA's RTW Model for Corrosion Control and Process Chemistry, can be used to calculate the CCPP based on water quality inputs. These can be useful tools for evaluating the impacts of various water quality or treatment changes on scaling potential.

If elevated levels of magnesium exist in the water, replacing calcium hardness with total hardness measurements will account for both calcium and magnesium precipitates. A CCPP greater than 10 mg/L as $CaCO_3$ (>4 mg/L as Ca) is undesirable because it may lead to hydraulic stress in distribution mains, reduce the life of residential hot-water heaters, and interfere with passivating pipe scales. Maximum precipitate levels may vary between utilities. The city of Tampa, Fla., established a lower maximum calcium precipitation concentration of 7 mg/L as $CaCO_3$ in the distribution system (Brodeur and Flourence 2007). At higher levels, the city was observing a high number of service pump failures and inaccuracies in residential meters.

It is important to note that groundwater systems may contain elevated carbon dioxide levels compared to surface water systems. When running bench tests, open beakers are acceptable for surface water sources and groundwater sources that are exposed to the atmosphere prior to pH adjustment. However, groundwater systems that do not contain atmospheric exposure (i.e., pump to distribution or use hydropneumatic tanks) prior to pH adjustment should use glass zero-headspace containers to prevent the gas from escaping. Carbon dioxide (carbonic acid) may be a factor in calcium carbonate precipitation in groundwater systems where the pH adjustment chemical is injected into the line before the water is exposed to atmospheric pressures.

Carbon dioxide (carbonic acid in water) will convert into carbonate as the pH increases. This conversion requires more pH chemical and increases alkalinity and DIC as the pH continues to increase. In this situation, calcium carbonate may precipitate at a lower pH than it would under atmospheric conditions.

CCPP can change in surface water systems on a seasonal basis. Therefore, bench testing should be conducted under source water conditions containing the highest DIC levels.

As an example of a calcium carbonate precipitation bench test, consider a surface water utility that needs to raise the entry-point pH from 7.5 to 8.3. In addition, the current soda ash feed will be replaced with caustic soda. Desktop models predict a CCPP of 1 mg/L as $CaCO_3$. The operator conducts a bench test to verify the model by vigorously mixing caustic soda to filter effluent water until the pH increases to 8.3. Once the target pH is reached, mixing is reduced to a very slow velocity for 24 hours. After mixing is

Courtesy of DC Water

Figure 4-7 Apparatus setup for filtering calcium carbonate and other precipitates such as iron, manganese, and phosphate

complete, the operator measures the amount of calcium hardness in the jar and obtains a reading of 100 mg/L as $CaCO_3$. Approximately 100 mL of water is filtered through a 0.45-µm filter (Figure 4-7). The operator proceeds to measure the calcium hardness level of the filtered water and obtains a reading of 96 mg/L as $CaCO_3$. Subtracting the dissolved calcium from the total calcium (100 mg/L as $CaCO_3$ – 96 mg/L as $CaCO_3$), a calculated value of 4 mg/L as $CaCO_3$ is obtained. Results were higher than predicted by the model; however, the precipitation potential should not cause a reduction in hydraulics, although calcium carbonate is expected to moderately accumulate in dead-end or low-flow areas. The utility may want to further study dead-end precipitation effects by piloting an isolated section of the distribution system.

Oxidation–Reduction Potential Test

The ORP test is typically conducted when a utility is changing disinfectants or changing concentrations of the same disinfectant. The purpose of the test is to measure current oxidation potentials and compare them to the oxidation potentials that occur because of a change in concentration of the current disinfectant or a change to an entirely different disinfectant. Changes in ORPs greater than 100 mV may warrant additional studies or construction of pipe loops to evaluate the full effect of the disinfectant change.

To run an ORP test, fill one 1-L beaker with finished water (entry point) and another 1-L beaker with baseline water (*baseline* refers to treated water collected immediately before secondary disinfection is added or collected immediately before primary disinfection is added if there is no secondary disinfection treatment). To the baseline beaker, add the new disinfectant at the designated concentration; or if continuing with the same disinfectant, add the new concentration. Using an ORP meter, measure ORP in each beaker using Standard Method 2580 for ORP (APHA et al. 2012).

Phosphate Test

A water system planning to add a phosphate-based chemical to the water may want to conduct bench tests to determine the amount of reactive phosphate that will remain available for passivation after it is added to the water. Phosphate can react with dissolved aluminum and iron to form phosphate precipitates. In addition, water with saturated phosphate levels (>3.0 mg/L as PO_4) can form precipitates with ferrous and calcium compounds (Tesfai et al. 2006) found in the distribution system. If using a polyphosphate or blended phosphate, bench tests might provide information relating to the breakdown of polyphosphate to orthophosphate. If phosphates are used for sequestration and corrosion purposes, bench tests can also identify time release of iron and manganese.

Steps for running a phosphate bench test are as follows:

1. Using finished water, add the desired phosphate concentrations to a 1-L jar.
2. Mix rapidly for 30 seconds and then continue to mix at a slow rate for 24 to 48 hours or to the longest detention time in the system.
3. Measure the orthophosphate.
4. Filter approximately 100 mL of sample through a 0.45-μm filter.
5. Measure the filtered orthophosphate level. This amount is the remaining phosphate that is available for corrosion control.
6. Subtract the filtered orthophosphate concentration from the unfiltered orthophosphate concentration. This amount is the precipitated phosphate. Precipitated phosphate levels greater than 1.0 mg/L as PO_4 might warrant pilot testing in the distribution system.
7. If using blended or polyphosphate for sequestration purposes, orthophosphate readings should be taken at the start of the test as well as at the end. An increase in unfiltered orthophosphate will reflect the amount of polyphosphate that was converted to orthophosphate. In addition, measurements for iron and/or manganese should be included in steps 3 through 6. When polyphosphate breaks down, it can release bound iron or manganese, which when exposed to a disinfectant will become an oxidized precipitate. Precipitated iron levels greater than 0.3 mg/L and precipitated manganese levels greater than 0.05 mg/L may warrant pilot testing.

COUPON AND PIPE LOOP TESTS

Coupon and/or pipe loop tests can also be used to help with the selection of a corrosion control treatment method or to optimize existing treatment. Details on these types of testing are provided in chapter 6.

CONCLUSION

Selection of the best corrosion technology is dependent on the intended treatment objectives, a thorough understanding of the characteristics of the water being treated, and the properties of the technology/products available. In some cases, several treatment options will be available. Consideration will also need to be given to impacts of treatment chemicals on wastewater discharge limits or concentrations of metals in sludge. Table 4-6 provides a summary of criteria for selecting a corrosion control technology to evaluate.

The complete elimination of corrosion is almost impossible. Technologies exist to reduce or inhibit corrosion, but the feasibility of each technology will depend on the economics of an individual system and the risk/aesthetic trade-offs. The success of a particular corrosion control treatment is dependent on the specific water quality and piping of an individual system. A particular method may be successful in one system and not in another. Bench tests can provide purposeful information relating to potential reactions with baseline water when considering whether to add a new chemical for corrosion control treatment. However, they cannot provide information relating to how the water reacts with pipe scales such as metals release, scale stability, and biofilm formation.

Table 4-6 Corrosion control technique selection criteria

Criteria	Selection
Effectiveness	How effective is each treatment in reducing standing metal concentrations? How closely will the reductions approach regulatory levels? How consistent are the standing metal concentrations? Has it reduced pitting corrosion?
Secondary impacts	What secondary impacts are associated with each treatment? • THM levels • HAA levels • Disinfection efficiency • Corrosion by-products • Iron and manganese precipitation • Compatibility with existing water quality • Compatibility with needs of large-volume users or critical industries • Compatibility with wastewater standards and capabilities
Operability	How easy is the treatment technology to use and how safe will it be to operate? • Liquid versus powder • Staffing • Training • Maintenance
Reliability	How reliable is the technology for each treatment approach?
Flexibility	How flexible will each treatment be to future water demands? Is the technology responsive to operational changes? • Future planning • Future regulatory issues
Costs • Capital • Operations and maintenance • Chemicals	How does the present worth value of costs compare between treatments? • Inflation • Life cycle • Interest rates
Implementation factors • Ease of implementation • Time to implement • Public opinion	Will there be legal or regulatory ramifications to implementing the treatment? How long will it take to design and construct facilities for each treatment? How will the public respond to changes in water quality characteristics and any secondary impacts associated with each treatment?

REFERENCES

APHA, AWWA, and WEF (American Public Health Association, American Water Works Association, and Water Environment Federation). 2012. *Standard Methods for the Examination of Water and Wastewater*, 22nd ed. Edited by E.W. Rice, R.B. Baird, A.D. Eaton, and L.S. Clesceri. Washington, D.C.: APHA, AWWA, and WEF.

Arweiler, S., Y.K. Cohen, and C.G. Abernathy. 2003. Polyphosphates—The solution to distribution system low residuals, biofilm and pipe corrosion—Part II. In *Proceedings of the AWWA Water Quality Technology Conference*. Denver, Colo.: AWWA.

AWWA (American Water Works Association). 2004. *AWWA Water Operator Field Guide*. Denver, Colo.: AWWA.

AWWA. 2005. *Managing Change and Unintended Consequences: Lead and Copper Rule Corrosion Control Treatment*. Denver, Colo.: AWWA.

AwwaRF and TZW (Awwa Research Foundation and DVGW-Technologiezentrum Wasser). 1996. *Internal Corrosion of Water Distribution Systems*, 2nd ed. Denver, Colo.: AwwaRF.

Benjamin, L., R.W. Green, A. Smith, and S. Summerer. 1992. Pilot testing a limestone contactor in British Columbia. *Journal AWWA*, 84(5):70–79.

Benjamin, M.M., S.H. Reiber, J.F. Ferguson, and E. Vanderwerff. 1990. *Chemistry of Corrosion Inhibitors in Potable Water*. Denver, Colo.: AwwaRF.

Boulay, N., and M. Edwards. 2001. Role of temperature, chlorine, and organic matter in copper corrosion by-product release in soft water. *Water Research*, 35(3):683–690.

Brodeur, T., and R. Flourence. 2007. Project seeks to balance corrosion control, scale formation. *Water World*, November.

Brown, R., N. McTigue, and D. Cornwell. 2013. Strategies for assessing optimized corrosion control treatment of lead and copper. *Journal AWWA*, 105(5):62–74.

Cantor, A.F., D. Denig-Chakroff, R.R. Vela, M.G. Oleinik, and D.L. Lynch. 2000. Use of polyphosphate in corrosion control. *Journal AWWA*, 92(2):95.

Cardew, P.T. 2009. Measuring the benefit of orthophosphate treatment on lead in drinking water. *Journal of Water and Health*, 7(1):123–131.

Clesceri, N.L., and G.F. Lee. 1965. Hydrolysis of condensed phosphates—II: Sterile environment. *International Journal of Air and Water Pollution*, 9:743.

Dodrill, D.M., and M. Edwards. 1995. Corrosion control on the basis of utility experience. *Journal AWWA*, 87(7):74–85.

Douglas, I., J. Guthmann, Q. Muylwyk, and V. Snoeyink. 2004. Corrosion control in the city of Ottawa—Comparison of alternatives and case study for lead reduction in drinking water. In *Proceedings of the 11th Canadian National Drinking Water Conference and 2nd Policy Forum*, Calgary, Alberta, April 3–6.

Dyksen, J.E., J.A. Clement, J. Brandt-Edwards, R. Hoehn, and C. Spencer. 2007. *Long-Term Effects of Disinfection Changes on Water Quality*. Denver, Colo.: AwwaRF.

Edwards, M., and L.S. McNeil. 2002. Effect of phosphate inhibitors on lead release from pipes. *Journal AWWA*, 94(1):79–90.

Edwards, M., L. McNeill, T. Holm, and M. Lawrence. 2001. *Role of Phosphate Inhibitors in Mitigating Lead and Copper Corrosion*. Denver, Colo.: AwwaRF and AWWA.

Frommell, D.M., C.M. Feld, V.L. Snoeyink, B. Melcher, and C. Feizoulof. 2004. Aluminum residual control using orthophosphate. *Journal AWWA*, 96(9):99–109.

Hayes, C.R., S. Incledion, and M. Balch. 2008. Experience in Wales (UK) of the optimisation of ortho-phosphate dosing for controlling lead in drinking water. *Journal of Water and Health*, 6(2):177–185.

Health Canada. 2008. *Guidelines for Canadian Drinking Water Quality: Guideline Technical Document—Chlorite and Chlorate*. Ottawa, Ont.: Health Canada, Water Quality and Health Bureau, Healthy Environments and Consumer Safety Branch.

Health Canada. 2009. *Guidance on Controlling Corrosion in Drinking Water Distribution Systems*. Ottawa, Ont.: Health Canada.

Hill, C.P., and Cantor, A.F. 2011. Manual M28. *Internal Corrosion Control in Water Distribution Systems*, 1st ed. Denver, Colo.: AWWA.

Holm, T.R., and M. Edwards. 2003. Metaphosphate reversion in laboratory and pipe-rig experiments. *Journal AWWA*, 95(4):172–178.

Holm, T.R., and M.R. Schock. 1991. Potential effects of polyphosphate products on lead solubility in plumbing systems. *Journal AWWA*, 83(7):76.

Hozalski, R.M., Y. Tan, and X. Dai. 2010. *An Investigation of Stannous Chloride as an Inhibitor of Lead Corrosion*. Denver, Colo.: Water Research Foundation.

Jalil, A.M. 2003. Limestone Contactors as a Corrosion Control Technology for Small Public Water Systems: An Educational Module for Drinking Water Professionals. Master's thesis. University of Tennessee. Available at: http://trace.tennessee.edu/utk_gradthes/2021

Kirmeyer, G.J., G. Pierson, J. Clement, A. Sandvig, V. Snoeyink, W. Kriven, and A. Camper. 2000. *Distribution System Water Quality Changes Following Corrosion Control Strategies*. Denver, Colo.: AwwaRF and AWWA.

LeChevallier, M.W., N.J. Welch, and D.B. Smith. 1996. Full-scale studies of factors related to coliform regrowth in drinking water. *Applied and Environmental Microbiology*, 62:2201–2211.

Leitch, R. 1996. Corrosion control: Begin with pH adjustment. *Water-Engineering and Management*, 143(1):22–27.

Letterman, R.D., and S. Kathari. 1996. A computer program for the design of limestone contactors. *Journal of the New England Water Works Association*, 110(1):42–47.

Letterman, R.D., C.T. Driscoll Jr., M. Haddad, and H.A. Hsu. 1986. *Limestone Bed Contactors for Control of Corrosion at Small Water Utilities*. Research and Development Report, EPA/600/S2-86/099. Cincinnati, Ohio: USEPA, Water Engineering Research Laboratory.

Letterman, R.D., M. Hadad, and C.T. Driscoll. 1991. Limestone contactors: Steady-state design relationship. *Journal of Environmental Engineering-ASCE*, 117(3):339–358.

Lowther, E.D., and R.H. Moser. 1984. Detecting and eliminating coliform regrowth. In *Proceedings of the Water Quality Technology Conference*. Denver, Colo.: AWWA.

Lytle, D.A., and M.R. Schock. 2005. Formation of Pb(IV) oxides in chlorinated water. *Journal AWWA*, 97(11):102–114.

Lytle, D.A., M.R. Schock, J.A. Clement, and C.M. Spencer. 1998. Using aeration for corrosion control. *Journal AWWA*, 90(3):74–88.

Lytle, D.A., P. Sarin, and V.L. Snoeyink. 2003. The effect of chloride and orthophosphate on the release of iron from drinking water distribution system cast iron pipe. In *Proceedings of the AWWA Water Quality Technology Conference*. Denver, Colo.: AWWA.

McNeill, L.S., and M. Edwards. 2000. Phosphate inhibitors and red water in stagnant iron pipes. *Journal of Environmental Engineering*, 126(12):1096–1102.

McNeill, L.S., and M. Edwards. 2001. Iron pipe corrosion in distribution systems. *Journal AWWA*, 93(7):88–100.

McNeill, L.S., and M. Edwards. 2002. Phosphate inhibitor use at US Utilities. *Journal AWWA*, 94(7):57.

Reiber, S. 2004. Disinfection Byproducts vs. Corrosion: A Case Study on the DC Water Lead Experience. Presentation to USEPA, April 2004.

Reiber, S. 2006. Corrosion, red water, and distribution system water quality. In *Proceedings of the AWWA Annual Conference and Exposition*. Denver, Colo.: AWWA.

Reiber, S., S. Edwards, S. Poulson, S. Perry, S. Patel, and D. Dodrill. 1997. *A General Framework for Corrosion Control Based on Utility Experience*. Denver, Colo.: AwwaRF and AWWA.

Rezania, L. April 2015. Personal communication.

Robinson, R.B., R.A. Minear, and J.M. Holden. 1987. Effects of several ions on iron treatment by sodium silicate and hypochlorite. *Journal AWWA*, 79(7):116.

Rushing, J., L. McNeill, and M. Edwards. 2002. Some effects of aqueous silica on the corrosion of iron. *Water Research*, 37(5):1080–1090.

Schock, M.R., J. Clement, D.A. Lytle, A.M. Sandvig, and S.M. Harmon. 1998. Replacing polyphosphate with silicate to solve problems with lead, copper and source water iron. In *Proceedings of the AWWA Water Quality Technology Conference*. Denver, Colo.: AWWA.

Schock, M.R., D.A. Lytle, A.M. Sandvig, J. Clement, and S.M. Harmon. 2005. Replacing polyphosphate with silicate to solve lead, copper and source water iron problems. *Journal AWWA*, 97(11):84.

Spencer, C.M. 1998. Aeration and limestone contact for radon removal and corrosion control. *Journal of the New England Water Works Association*, 112(1):60–69.

Stericker, W. 1945. Protection of small water systems from corrosion. *Industrial Engineering Chemistry*, 37:716.

Tesfai, F., P. Constant, S. Reiber, R. Giani, and M. Donnelly. 2006. Precipitate formation in the distribution system following orthophosphate addition. In *Proceedings of the AWWA Water Quality Technology Conference*. Denver, Colo.: AWWA.

USEPA (US Environmental Protection Agency). 1986. *Limestone Bed Contactors for Control of Corrosion at Small Water Utilities*. EPA/600/2-86/099. Cincinnati, Ohio: USEPA.

USEPA. 2016. *Optimal Corrosion Control Treatment Evaluation Technical Recommendations for Primacy Agencies and Public Water Systems*. EPA 816-B-16-003. Washington, D.C.: USEPA, Office of Water.

Videla, H. 2002. Prevention and control of biocorrosion. *International Biodeterioration and Degradation*, 49:259–270.

Vik, E.A., R.A, Ryder, I. Wagner, and J.F. Ferguson. 1996. Mitigation of corrosion effects. In *Internal Control of Water Distribution Systems*. New York: McGraw-Hill.

Wasserstrom, L.W., M.R. Schock, and S.A. Miller. 2015. Practical implications from observed lead pipe scale mineralogy in a blended phosphate treated system. In *Proceedings of the AWWA Annual Conference and Exposition*. Denver, Colo.: AWWA.

WHO (World Health Organization). 2011. *Guidelines for Drinking-Water Quality*, 4th ed. Geneva: WHO.

Williams, S.M. 1990. The use of sodium silicate and sodium polyphosphate to control water problems. *Water Supply*, 8:195.

This page intentionally blank.

Chapter 5

Distribution System Corrosion Control: Design, Operations, and Maintenance Techniques

Richard Giani, CH2M
Richard Brown, Cornwell Engineering Group
Abigail F. Cantor, Process Research Solutions, LLC
Andrew Hill, Confluence Engineering Group, LLC
Brian Lakin, McMillen Jacobs Associates
Stephen J. Randtke, University of Kansas
Vanessa Speight, *Stewart Husband*, and *Joby Boxall*, University of Sheffield
Elizabeth Turner, North Texas Municipal Water District

INTRODUCTION

Implementing and maintaining optimum corrosion control requires appropriate water quality leaving the treatment plant and the ability to maintain that quality throughout the distribution system. Many factors can alter the water quality and make the distribution system more susceptible to corrosion, such as long detention times, tuberculated pipe scales that can provide a haven for microbes that can alter the chemistry, too high or low

Table 5-1 Chapter 5 key points

- Distribution corrosion can be minimized during system upgrades by considering system design with low water age, optimal flow, proper storage tank mixing, and pipe materials resistant to corrosion.
- Routine system maintenance such as flushing, cleaning, and lining of mains can reduce corrosion and/or remove the by-products of corrosion.
- Techniques for controlling microbiologically influenced corrosion (MIC) include reducing accumulation of metal particles within biofilms, involving management of disinfection, control of water age, hydraulic mobilization of biofilm, and routine main cleaning.
- Use of tank mixers, automatic flushing systems, and a grid/loop distribution system design can minimize stagnation and reduce water age.
- The design of pipes and structures in the distribution system can be as important as the selection of materials.
- Maintaining disinfectant residual reduces MIC and maintains a stable oxidation potential that can stabilize pipe scales.
- Routine flushing of mains via unidirectional flushing techniques can effectively remove biofilm, and built-up sediment and tuberculation, but could also lead to exposure of less stable scales, which could cause post-cleaning water discoloration after flushing is complete.
- Cleaning the pipeline with procedures such as pigging, or mechanical scrapers, can effectively remove loose sediment and weak, exposed tubercles without causing the potential for post-cleaning discoloration.
- Controlling shear stress can minimize sediment deposit and slowly remove pipe tubercles over time without causing noticeable water discoloration.
- When replacing lead service lines, critical steps must be taken before, during, and after the process to minimize scale disturbance and ensure public protection.

a velocity, and so forth. The most common ways to minimize corrosion in the distribution system are

- distribution system design considerations,
- adjustment of water chemistry,
- short detention times,
- proper routine maintenance of the distribution system,
- linings and coatings, and
- pipe replacement with appropriate material and procedures.

Key points covered in this chapter are listed in Table 5-1.

DISTRIBUTION SYSTEM DESIGN CONSIDERATIONS

A water utility can minimize distribution corrosion by properly selecting distribution system materials and having a good engineering design. The reality is, however, that most utilities are forced to deal with existing distribution systems that were designed with little regard for corrosion. Nonetheless, utilities can take advantage of system design for corrosion control during system upgrades, planned system materials replacement, and expansion of the distribution system.

System and Pipe Materials

The piping materials available for use in a distribution system and premise plumbing include iron, cement, steel, lead, copper, brass, and plastic. Water mains installed between the late 1800s and World War II are usually unlined cast iron (i.e., gray iron) (AWWA 2010b). Cement-mortar-lined ductile-iron pipe began to be installed in the 1950s. Polyvinyl

Table 5-2 Corrosion properties of materials frequently used in water distribution systems

Plumbing Material	Corrosion Resistance	Primary Contaminant From Pipe
Asbestos–cement, concrete, and cement linings	Good corrosion resistance. Immune to electrolysis. Aggressive waters can leach calcium from the cement. Polyphosphate sequestering agents can deplete the calcium and substantially soften the pipe.	Asbestos fibers (from erosion), increase in pH, aluminum, and calcium. Trace metals (cadmium, chromium, barium, and aluminum) due to presence in some cements.
Brasses and bronzes	Good overall resistance. Different types of brass respond differently to water chemistry. Subject to dezincification by waters of pH >8.3 with high ratio of chloride-to-carbonate hardness.	Lead, copper, zinc, selenium, bismuth, and phosphorus.
Copper	Good overall corrosion resistance. Subject to corrosive attack from high flow velocities, soft water, chlorine, dissolved oxygen, low pH, and high inorganic carbon levels (alkalinity). Subject to microbiologically influenced corrosion. May be prone to pitting failures.	Copper and possibly iron, zinc, tin, antimony, arsenic, cadmium, and lead from associated pipes and solder.
Galvanized iron or steel	Subject to galvanic corrosion of zinc by aggressive waters, especially of low hardness. Corrosion is accelerated by contact with copper materials. Corrosion is accelerated at higher temperatures such as in hot water systems. Corrosion is affected by manufacturing process of the pipe and galvanized coating.	Zinc and iron. Cadmium, chromium, barium, aluminum, and lead are possible because of impurities in the galvanization process.
Iron, unlined cast or ductile	Can be subject to surface erosion by aggressive waters and tuberculation in poorly buffered waters.	Iron, resulting in red water, complaints, and turbidity.
Lead	Corrodes in soft water with pH <8 and in hard waters with high inorganic carbon levels and pH <7.5 or >8.5.	Lead
Plastic	Resistant to corrosion	Some pipes contain metals in plasticizers, notably lead. Plasticizers are an emerging contaminant issue.
Steel, mild	Subject to uniform corrosion. Affected primarily by high dissolved oxygen and chlorine levels and poorly buffered water.	Iron. Resulting in red water, complaints, and turbidity.

chloride (PVC) pipes were introduced in the 1970s followed by high-density polyethylene (HDPE) in the 1990s. Asbestos–cement pipe was used from the 1930s but is no longer in use due to the hazards of asbestos (AWWA 2010b).

The selection of piping material should be based on water chemistry of the system and cost. The mechanical properties of the piping material should also influence the selection of materials. In addition, some jurisdictions regulate the components of the distribution system. For example, Naperville, Ill., requires its mains to be cement-lined ductile-iron pipe, with a minimum Class 52 thickness designation, and polyethylene encasement. It also requires that service lines 2 in. (5.08 cm) or less (<2 in.) inside diameter must be copper pipe (Type K), and all services 3 in. (7.62 cm) or greater (>3 in.) inside diameter must be Class 52 ductile-iron pipe with cement lining (City of Naperville 2002). When selecting materials to replace old or broken mains, putting new mains into service, or rehabilitating existing pipes, the utility should choose a material that will not corrode in the environment (external) or under the water quality conditions anticipated (internal). Table 5-2 will assist in the selection of piping materials.

Figure 5-1 Asbestos–cement pipe

Asbestos–cement pipe, cement–mortar linings, and concrete pipe. Asbestos–cement pipe (Figure 5-1) was once popular in water distribution systems because of its corrosion resistance. Health concerns from the potential release of asbestos fibers have limited the installation of new asbestos–cement pipe (AwwaRF and TZW 1996). Concrete pipe and cement-mortar-lined pipe are generally resistant to corrosion. Asbestos–cement pipe and prestressed concrete cylinder pipe behave very similarly with the exception of the possible release of asbestos from asbestos–cement pipes when that pipe deteriorates. The leaching of calcium and its complexes from cement materials raises the pH, which can reduce the effectiveness of free residual chlorine as a disinfectant, and can cause unwanted precipitation of minerals, thus resulting in cloudy or turbid water and poor taste.

The cement matrix can be quite complex. More than 100 compounds and phases important to the chemistry of portland cement and related cements have been described and identified (Schock 1989). The main components of portland cement are tricalcium silicate, dicalcium silicate, and tricalcium aluminate, plus small amounts of iron and magnesium.

The corrosion resistance of mortar linings is dependent on the mortar density. Mixture of cement and sand in a 1:1 ratio is more resistant to concrete corrosion than other ratios.

One of the best protective measures for all cement-based materials is calcium carbonate deposition, which can fill voids in the cement matrix and prevent leaching. Consequently, the integrity of cement-lined pipe does depend on such calculations as the calcium carbonate precipitation potential and the Langelier saturation index. Unfortunately for asbestos–cement pipes, calcium carbonate deposition still leaves asbestos fibers exposed and vulnerable to erosion. Iron coatings are beneficial by preventing exposure of the fibers at the pipe surface and reducing the possibility of release of asbestos fibers into the bulk water. However, they do not seal the cement pipe matrix from dissolution. Iron coatings may also have their own corrosion-related problems. Manganese coatings can provide a similar protection to that provided by iron. Silica may also be beneficial by assisting the formation and adsorption of iron colloids onto the pipe surface.

Zinc orthophosphate, zinc sulfate, and zinc chloride have been used in the lab to prevent softening of cement pipes. Strong sequestering agents such as polyphosphates can attack the pipe by dissolving calcium, aluminum, iron, and magnesium from the cement matrix.

Additional aspects of cement must be considered:

- Metal impurities in the cement mix can leach into the water (Guo et al. 1998, Berend and Trouwborst 1999).
- Other ions, such as sulfate, in the water in contact with the cement lining can affect its integrity by forming hydrated compounds that cause the cement to crack (AwwaRF and TZW 1996).

Figure 5-2 Brass faucet fittings and faucets can contain as much as 8 percent lead by weight

Figure 5-3 Bronze fitting typically used in premise plumbing

Brasses and bronzes. Brasses and bronzes are primarily used in home plumbing fixtures and fittings (Figures 5-2 and 5-3). *Brass* is the term used for alloys of copper and zinc. The amount of zinc may vary from 5 to 45 percent. Lead is often added to brass to improve machinability. *Bronze* refers to a broad range of copper alloys, usually with tin as the main additive but sometimes with other added elements such as phosphorus, manganese, aluminum, or silicon. Valves, faucets, and water meters may be composed of brass, bronze, or other materials. Most studies on corrosion of brass and bronze have focused on dezincification. The corrosion process can contribute lead, copper, and zinc to the water. The contribution of brass fixtures to first-draw lead and copper samples in tap water monitoring programs is now well recognized (Schock et al. 1996). Prior to 2013, brass fixtures contained as much as 8 percent lead.

In studies, polyphosphates proved to be effective in suppressing corrosion of brass and bronze as long as zinc was present at a concentration of 1 mg/L (Schock et al. 1996). Orthophosphates also appear to control corrosion.

Copper. Copper (Figure 5-4) is a common material for premise plumbing. Reasons for its use include ease of installation and corrosion resistance. Copper corrosion can result in pinhole leaks and blue or green water. Copper is much more resistant to uniform corrosion than the other metals typically found in potable water systems except in soft waters with low pH. Also, high-alkalinity groundwaters are particularly aggressive toward copper materials. Copper is prone to erosion corrosion because of high-velocity water, typically seen in recirculating hot water systems where water flows past pipe bends. Impingement attack of copper by high water velocities is one of the problems of using copper pipe (Schock 1999).

Figure 5-4 Copper pipe commonly used for service lines and premise plumbing

Figure 5-5 Galvanized pipe commonly found in homes built before 1980 and may be a source of lead through iron adsorption and release

Adjustment of pH is very successful in controlling copper corrosion and is usually the least expensive option. Anion concentrations (chloride, sulfate, bicarbonate, orthophosphate) are purported to exert an influence on corrosion.

Numerous articles can be found on the effect of phosphate to control copper release in real systems and from pipe-rig studies. As described in chapters 2 through 4, orthophosphate in sufficient dosage at the proper pH can reduce copper corrosion.

Galvanized steel. Galvanized steel pipe (Figure 5-5) was the dominant material used in premise plumbing but has been replaced in the past few decades by copper and plastic. In the United States and northern Europe, new installations of galvanized pipe are rare. Galvanized pipe may be found in homes built prior to the 1980s (AwwaRF and TZW 1996). Galvanized steel pipe consists of a base steel layer with layers of iron–zinc alloy that approach pure zinc at the interior surface of the pipe. Impurities in the iron–zinc alloy include lead, barium, chromium, and cadmium. Galvanized pipe is expected to give good service in hard or soft waters in the pH range of 7.0 to 8.5 but performs better in hard waters.

The corrosion issues with galvanized pipe are complex. When the pipe is new, corrosion products will be zinc and impurities of the galvanized layer, such as lead and cadmium. The quality of the zinc coating plays a major role in corrosion. Problems with the galvanization process can lead to pitting instead of uniform corrosion. Once the zinc layers have been corroded away, the pipe behaves as a black iron pipe. Old galvanized pipe in premise plumbing, downstream to lead sources such as lead service lines (LSLs), have been identified as sinks for lead accumulation (McFadden et al. 2011). During higher water

Figure 5-6 Inside view of a tuberculated, unlined cast-iron main

velocity or pressure changes, these scales can release concentrated lead and other metals that have been embedded in the pipe scale over time.

According to several studies, orthophosphate effectively controls zinc solubility and galvanized steel corrosion (Trussell and Wagner 1996). Silicates can also reduce galvanized pipe corrosion, but most studies have focused on hot water systems. Adjustment of pH between 7.0 and 8.5 is also a very effective treatment.

Iron. Iron is the most common material in the distribution system and is usually found as cast iron, ductile iron, and steel. The forms of iron-based materials are distinguished by their nonferrous components. The major difference is that steel has lower carbon and silicon content than either cast iron or ductile iron.

Unlined cast-iron mains (Figure 5-6) have been in use in the United States since the early 1800s. Lined ductile-iron mains have been installed since the 1950s but became more common in the 1970s. Ductile iron is much stronger and less prone for cracking compared to cast-iron mains. The lining in ductile-iron pipe is intended to prevent or minimize corrosion and is usually asphaltic coating over a cement lining. Distribution systems older than 50 years may have a large portion of their distribution mains as unlined cast iron. Corrosion of the iron-containing materials may be either general or localized. The mechanisms of iron corrosion are covered in chapter 2.

Although chlorine has the greatest negative effect on new pipe, its impact diminishes as the pipe surface is passivated by corrosion products. Once corrosion products have formed, chlorine can provide some protection from corrosion by reducing the formation of ferrous iron, which is more soluble than ferric iron. Iron corrosion may also be reduced through alkalinity and calcium adjustment or the addition of a corrosion inhibitor (Clement et al. 2002).

Lead. In addition to lead pipes, sources of lead in household plumbing include pipe jointing compounds, lead solder, and brass and bronze fixtures. The use of lead for service lines was prohibited by the Safe Drinking Water Act Amendments. This also restricted the lead content of solder to less than 0.2 percent and less than 8 percent in other materials in contact with drinking water. The National Plumbing Code of Canada (NPC) prohibited the use of lead for LSLs in 1975. Lead solder used in plumbing was limited to a lead content of 0.2 percent in 1986, and the 1990 version of the NPC officially prohibited lead solders from being used in new plumbing or in repairs to plumbing. Lead-containing brass fittings, faucets, and valves may contain up to 8 percent lead (Health Canada 2009).

On January 4, 2014, the Reduction of Lead in Drinking Water Act required wetted surfaces of pipes, pipe fittings, and plumbing fittings and fixtures to contain no more than 0.25 percent lead, as a weighted average. The NPC was amended in November 2013 to reference relevant standards with the 0.25-percent-weighted-average lead content requirements (NRCC 2013). There are approximately 6.1 million LSLs in the United States (Cornwell et

Figure 5-7 Harvested lead service lines

al. 2016). Many water utilities began systematic replacement of LSLs (Figure 5-7) with the promulgation of the Lead and Copper Rule in 1991.

Corrosion control of lead-bearing materials may be accomplished through (1) the physical removal of lead-containing materials and (2) water quality adjustment. The removal of LSLs is expensive. In addition, it is not always possible to remove the entire LSL or all lead-containing materials, and recent studies have shown a temporary increase in lead in some homes as a result of the disturbance of pipe scales downstream of the replaced pipe (AwwaRF and USEPA 2008, McFadden et al. 2011, Del Toral et al. 2013). The best option is usually to approach lead corrosion on both fronts—remove as much lead-containing material as economically and physically possible while adjusting water chemistry. Research has shown that fully removing an LSL is far more beneficial than partial removal (AwwaRF and USEPA 2008).

Plastic. Many service lines and some premise plumbing are being replaced with plastic (Figure 5-8). Plastic lines may be either polyvinyl chloride (PVC), high-density polyethylene (HDPE), chlorinated polyvinyl chloride (CPVC), or cross-linked polyethylene (PEX). Plastic is a good choice of material for use in areas with aggressive water. A few local codes restrict the use of plastic pipe. Plastic is resistant to corrosion but has some non-corrosion-related disadvantages. Plastic pipe is less resistant to internal biofilm growth. Buried plastic pipe can be difficult to locate because it does not conduct electrical current for tracing. Plastic pipe is also more susceptible to permeation by organic solvents that may be present in the ground.

Steel. Steel pipe has been used in the United States since 1852 (AWWA 2010b). It is typically used in situations where there is high-pressure or large-diameter transmission mains. Steel pipe walls are usually thin and are sensitive to corrosion on both the inside and outside of the pipe. Typically, steel pipes (after the 1950s) contain protective internal linings such as cement mortar or epoxy. The exterior of the pipe is usually provided with additional external protection if buried. Common external protection includes plastic coatings, bituminous materials, and polyethylene tape as well as cathodic protection. Older steel pipe, where the coatings no longer exist, are highly susceptible to corrosion, including galvanic corrosion.

Galvanic corrosion. It has been previously mentioned that compatible materials should be used to reduce galvanic corrosion. When two dissimilar metals are connected, galvanic corrosion can occur. The galvanic couples that commonly occur in the water and wastewater industries can be rated according to severity. Examples of couples that can cause severe galvanic corrosion in the water industry are listed in Table 5-3.

In addition, galvanic corrosion can occur within a service line. There has been an increased focus on galvanic corrosion because replacement of lead water service lines often includes a connection of copper pipe directly to lead pipe downstream or the use

Figure 5-8 Plastic water supply pipe

Table 5-3 Galvanic couples that can cause severe corrosion

Anodic Site	Cathodic Site
Dielectric-coated ductile-iron pipe	Mortar-coated steel pipe
Steel pipe (dielectric coated)	Copper services, blowoffs
Steel pipe (mortar coated)	Copper services, blowoffs
Polyethylene-encased ductile-iron pipe	Mortar-coated steel pipe

of a brass coupling connector. This occurs because most water service lines are owned or controlled from the water main to the curb stop by a municipal water utility, whereas the lead line from the curb stop to the building is owned by the property owner. Often the municipal line is replaced with copper while the property owner's line remains lead.

Contradicting studies on the significance of the galvanic corrosion that occurs at the copper–lead pipe connection were compared (Wang et al. 2013). Several studies reported that increasing the chloride-to-sulfate mass ratio in the water could accelerate lead release from galvanic corrosion. Several water quality parameters were also studied in terms of increasing or decreasing the copper–lead galvanic corrosion. However, other researchers agreed that the effects on the copper–lead galvanic corrosion were present when using new lead pipes. But older lead pipes with existing passivating scales did not show significant galvanic corrosion when coupled with copper. To reconcile these differences, researchers ran laboratory tests using new copper pipe connected to lead pipe harvested from a water distribution system (Wang et al. 2012). Three types of connections between the copper and lead were tested: a conventional brass coupling, a dielectric brass coupling with plastic bushing, and an all-plastic transition coupling. Varying levels of chloride-to-sulfate mass ratio were also tested. The lowest lead release was observed for the plastic-coupled pipes, as these types of pipes prevented galvanic corrosion from occurring. For the pipe combinations where galvanic corrosion occurred, the lead release was localized near the connection and diminished along the lead pipe away from the connection. In addition, increasing the chloride-to-sulfate mass ratio from 0.7 to 7 did not increase the amount of lead release.

A review of field data from system-wide partial LSL replacements in two large cities found that lead did increase initially after a galvanic connection of copper to lead water service lines but diminished over time, possibly within four to six months (Welter et al. 2013).

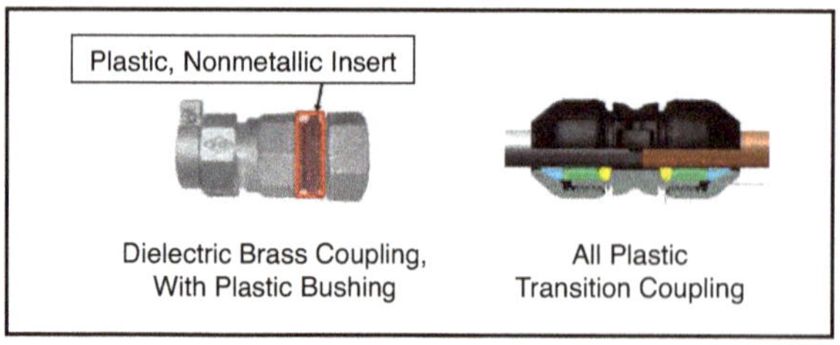

Source: *Welter et al. 2013*

Figure 5-9 Examples of couplings that can minimize galvanic corrosion on service lines

When connecting two dissimilar metals, insulators can be used to reduce the galvanic effect. The purpose of an insulator is to provide a nonconductive barrier between two conductive (metallic) components of a pipeline or piping system. The corrosion control philosophy and long-term integrity of many pipelines and piping systems depend on maintaining full metallic isolation at one or more locations over the life of the equipment. Insulator sites need to be selected early in the project design so that provisions for their installation, testing, and maintenance access may be properly incorporated.

The most common insulators used in the water industry are (DeCarlo 2004)

- insulating flange kits,
- casing insulators (centralizers),
- dielectric unions,
- insulating bushings,
- structural design (pipe penetrations through concrete),
- nonmetallic or link-seal–type sleeves,
- nonmetallic pipe spools,
- dielectric tubing connectors,
- direct current decoupling devices,
- insulating couplings, and
- monolithic insulators.

Each galvanic insulator has advantages and disadvantages depending on its application. Some municipal codes regulate the use of insulators in water systems (Figure 5-9). For example, alternative grounding methods for household wiring systems may be required if nonmetallic couplings are used on a service line that is less than 10 ft (3 m) from the house.

System Layout

The design and layout of pipes and structures in the distribution system can be as important as the selection of materials. Some important design considerations include the following:

- Avoid dead ends and stagnant areas, as this can increase particulate accumulation and biological activity.
- Plan for asset types and locations that support effective main cleaning.

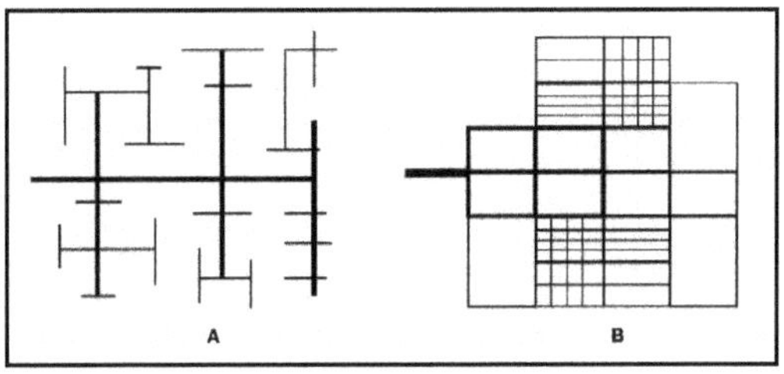

Source: *NRC 2006*

Figure 5-10 Branched (A) and grid/loop (B) designs

- Provide facilities for adequate flushing and cleaning of the system to minimize particulate accumulation and a consistent water quality that is conducive for passivation.
- Select an appropriate flow velocity (5–7 ft/s [1.52–2.13 m/s] at maximum system demand).
- Reduce mechanical stresses to avoid flexing of pipes and water hammer, as this can disrupt pipe scales.
- Reduce sharp turns and elbows to avoid erosion corrosion.
- Install flow and water quality monitoring equipment or provide access points for temporary monitors or sample collection.

The two basic configurations for most distribution systems are the branch and the grid/loop (AWWA 2010b) (Figure 5-10). In a branch system, smaller pipes branch off of larger pipes throughout the service area such that water can only take one path from source to consumer. In a grid/loop system, water may take several pathways from source to consumer. This design can be found in urban areas. Looping reduces some of the problems associated with stagnant water, such as adverse reactions with the pipe wall. But long loops can equally result in areas where water stagnates at the "tidal" points. Most systems will be a combination of the branched and looped designs.

The purpose of a distribution system is to deliver water to all customers of the system in sufficient quantity for potable drinking water and fire protection. Although only a small percentage of water is used for firefighting purposes, the sizing of water mains is partially based on fire protection requirements to ensure sufficient flow for a minimum period of time. Generally, 75 percent of the capacity of a typical drinking water system is devoted to firefighting (NRC 2006). Consequently, there may be long residence times before the water reaches the consumer, increasing the available time for reactions between the pipe wall and the water in the pipe as well as creating an environment to encourage the growth of microorganisms. In the absence of smaller distribution systems, water utilities have to implement flushing programs and increase dosages of disinfectants to maintain a residual at the ends of the distribution system. One suggested alternative is to have dual distribution systems—one dedicated to potable water use and the other to firefighting. The Irvine Ranch Water District in Irvine, Calif., is one of the most widely known utilities using a dual distribution system. Dual systems are most advantageous in new communities.

Accommodating flushing and cleaning. Most distribution system mains will require and benefit from the application of a controlled cleaning process one or more times during their service life. Implementation of main cleaning relies on the use of numerous distribution system assets (e.g., fire hydrants, valves), and the feasibility, effectiveness, and costs of a particular cleaning approach can be affected by asset type, size, and location. Therefore, utilities should anticipate the need for main cleaning during system layout planning and design and incorporate appropriate features to accommodate the anticipated cleaning methods. Failure to plan for this can result in costly future system retrofits or constrain the process to less effective methods. Notably, although anticipation of cleaning needs is an important planning consideration, in most cases involving new main installations that use modern materials and linings, rehabilitation is less likely such that it is generally not practical to plan for potential future rehabilitation. Thus, this section focuses on system design considerations to accommodate stand-alone main cleaning and flushing methods.

Several main cleaning options are available to utilities, as discussed later in this chapter. Common to each option is the need to plan and develop sequenced cleaning routes. A cleaning route consists of a discrete stretch of main with defined end points and targeted valve closures to ensure a unidirectional cleaning path. Route end points consist of a fire hydrant, blowoff assembly, or pipe access station. These serve as locations for pipe access and discharge of the process residual stream. Multiple cleaning routes are strung together in series to allow a controlled cleaning progression from an initial clean water interface throughout the targeted portion of the distribution system.

The development and use of efficient main cleaning routes are critical to the long-term success and cost-effectiveness of a cleaning program. Efficient routes are characterized not only by their ability to support the operational aspects of the preferred cleaning method but also their use of optimization criteria to accelerate field activities and reduce costs. On average, valve operations and related travel account for roughly 50 percent of field crew time for unidirectional flushing programs (Hill 2001), such that the incorporation of system design elements to reduce valve tracking and travel presents the most significant opportunity to reduce main cleaning operations and maintenance and life-cycle costs. Utilities should evaluate their system design standards and ensure that the placement and spacing of line valves, hydrants, blowoffs, and so forth can accommodate efficient and flexible route development while still meeting basic needs pertaining to system reliability, maintenance, and fire suppression. Table 5-4 provides general system design strategies and criteria that enhance route development in utility flushing and cleaning programs. Several of these concepts are illustrated in Figure 5-11.

In most instances, application of the aforementioned general system design elements will provide sufficient flexibility to create and optimize cleaning routes as needed at some future time. However, foam swabbing (also known as poly-pigging) presents unique infrastructure requirements because of the need to launch, navigate, and retrieve a solid foam device. Limitations on the physical compressibility of swabs and their inability to pass certain valves require special attention to system components and sizes along the cleaning path. Utilities should evaluate the potential future need for swabbing and, where anticipated, incorporate the additional system considerations detailed in Table 5-5. Utilities should develop and document the swab cleaning routes during the system design process to ensure that suitable components are installed and appropriately located.

Managing hydraulic retention time. Controlling the flow of water throughout the distribution system is a key factor in maintaining good water quality. The benefits of reducing hydraulic retention in the distribution system include reduction of disinfectant residual loss, reduction in nitrification, prevention of excess biofilm formation, and temperature control, all of which can have an impact on corrosion. The management of hydraulic retention time is a function of both the system design, discussed here, and the system operation, which is discussed in the next section.

Hydraulic retention time, which is sometimes called water age, is influenced by the design of the distribution system in two key areas: (1) the capacity of pipes relative to the demand and (2) the volume of water storage (AWWA 2010b). It is not uncommon for pipes to be sized for distant future demands and high fire flows, which results in low water velocities under typical demand conditions. These low velocities increase the retention time for water in the system. The impact of selecting a large pipe diameter can be seen in

Table 5-4 General system design strategies to support cleaning loop development

- Fire hydrants with a 5¼-in. main valve opening and 4½-in. or larger nozzle are typically needed to support flushing and swabbing operations. Smaller or non-standard hydrants and blowoff assemblies can typically be used to support air scouring or ice pigging (discussed later in this chapter), except where main sizes are larger than 12 in.
- Hydrant spacing and locations should conform to local standards and codes for fire suppression coverage, ease of accessibility, proximity to structures, and so on. In certain cases, the use of additional hydrants may enhance route development. For example, it may be beneficial to install additional hydrants near locations that are best suited for accommodating discharge residuals handling and disposal needs.
- Hydrants (or pipe access stations) should be installed wherever the nominal diameter of the main changes, on the upstream side of pressure-reducing valves, at actual or artificial dead ends (e.g., closed valves at pressure zone boundaries), and at low-elevation locations.
- Line valves should be co-located with hydrants and pipe access stations and placed where they are easily seen and readily accessible.
- Line valves should be installed immediately upstream and downstream of hydrants (and pipe access stations) located along a pipe span or loop, as well as on each branch of pipe tees and intersections in proximity to the junction point.
- Dead ends should be avoided to the extent possible. Where unavoidable, a fire hydrant or blowoff assembly should be installed at the end of the dead end. Blowoff assemblies should be avoided (in favor of hydrants) for mains that are 8-in. diameter or larger.
- For low-pressure areas, pipe gridding and looping should be maximized to provide an opportunity to reduce head loss leading up to designated cleaning routes.
- For areas with extensive topographical changes, loop development and asset placement should support the preferred flow direction of the specific cleaning technique. Swabbing and air scouring are performed in an uphill direction. Flushing and ice pigging can be performed in either direction, although downhill operation is preferred and may be needed to ensure adequate pressure to support targeted flow rates at the discharge hydrant.
- An identification (ID) scheme should be developed for all assets and asset ID tags included on electronic and hard-copy system maps.

Courtesy of Confluence Engineering Group, LLC

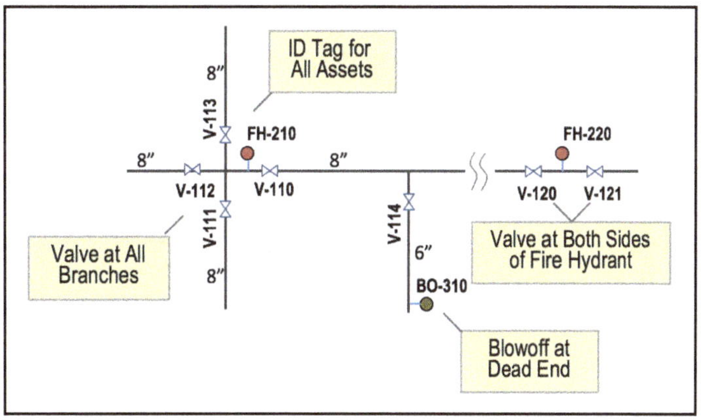

BO = blowoff; FH = fire hydrant; V = valve

Courtesy of Confluence Engineering Group, LLC

Figure 5-11 System design concepts to support main cleaning

Table 5-5 System design strategies to support foam swabbing

For pipes 8-in. diameter and smaller:

- Foam swabs can be compressed, launched, and retrieved through standard wet- and dry-barrel hydrants (minimum main valve opening of 5¼ in.) that are disassembled and gutted. For pipes 6 in. or smaller, swabs can also be retrieved through standard 2-in. blowoffs. Alternatively, utilities can install dedicated pipe access stations to support swabbing of smaller mains, although this is relatively uncommon owing to its high capital cost.
- Utilities should define the hydrant locations that will serve as route end points and specify a suitable hydrant type.
- Gate-style line valves should be installed in the main immediately upstream and downstream of these hydrants to ensure they can be used for both launch and retrieval. Butterfly valves should not be used because they will block the swab passage.

For pipes larger than 8-in. diameter:

- Pipe access stations are required at the cleaning route end points for swab launch and retrieval.
- A design standard should be developed that supports both launch and retrieval operations. A typical concept involves installation of an equal tee or wye in the main. The branch line should include an isolation gate valve and be capped with either a blind flange or a dedicated launch barrel. (See Figure 5-12 for an example of this setup.)
- For security and protection, these stations are typically housed below grade in a utility vault with a locked cover.

Other swabbing considerations:

- Owing to the relatively high capital cost associated with installation of pipe access stations, it is desirable to develop long swabbing route lengths (subject to other constraints, such as ensuring a suitable disposal location) to reduce the number of stations required for a particular area. Typical swabbing route lengths are 2,000 to 4,000 ft (610 to 1,220 m).
- It should be confirmed that there are no butterfly valves along the cleaning route. These valves prevent the passage of swabs. For large-diameter transmission mains, the location of any butterfly valves should be carefully considered relative to anticipated swabbing routes.

Source: *AWWA 2014c*

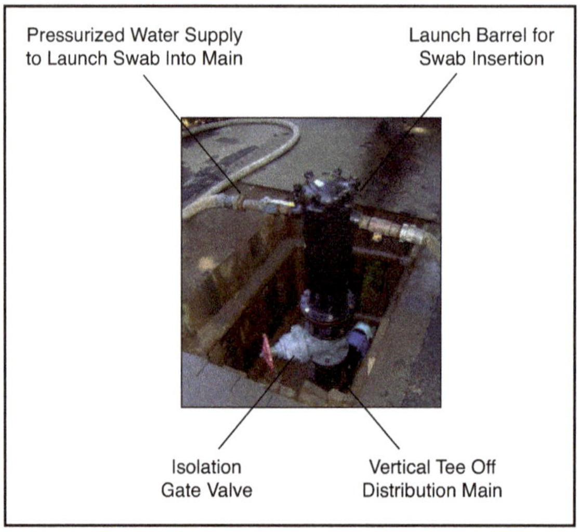

Courtesy of South Central Connecticut Regional Water Authority

Figure 5-12 Dedicated pipe access station for swab launch or retrieval

Table 5-6 Volumetric effect of pipe diameter

Pipe Diameter vs. Pipe Volume (per mile)							
	Pipe Diameter						
	2″	4″	6″	8″	10″	12″	18″
Gallons per mile	862	3,466	7,755	13,786	21,540	31,019	69,792
Liters per kilometer	2,028	8,153	18,241	32,427	50,665	72,961	164,161

Source: *LeChevallier et al. 2011*

Table 5-6, which lists the volume of water per mile of pipe for different distribution system pipe diameters. An increase in pipe diameter from 6-in. to 8-in. results in an additional 6,000 gallons of water per mile that is effectively stored in the piping. Therefore, to manage retention time, distribution system designs should use the lowest feasible pipe diameter that meets hydraulic requirements.

Although storage of water in the distribution system is necessary to maintain pressure, manage peak demands, and provide a backup supply for emergencies, excess storage volume can lead to increased hydraulic retention time system-wide. Reducing storage volume may not be possible because of regulatory requirements, but managing water age within storage tanks can be done. Good tank turnover will refresh the volume of water, ideally every 24 to 48 hours, to ensure that stagnation does not occur. Tank turnover is especially important when water temperature increases above 60°F (approximately 15°C), which favors accelerated microbial activity (LeChevallier et al. 2011).

Design of tanks should take mixing into consideration through sizing and positioning of inlet and outlet piping. Several products to promote tank mixing are also commercially available and should be selected based on the site-specific configuration and operational criteria.

Another option for hydraulic retention time management is to artificially increase water use through the use of automated flushing stations (AFSs). These stations utilize prefabricated, programmable devices to achieve localized flushing without the need for onsite crews. They provide a controlled low-rate conventional flush (flow is user specified, but typically less than 150 gpm) with the objective of improving circulation in areas of the system that experience bulk water quality degradation as a result of excessive water age. Although not designed for main cleaning, these devices can help stabilize water chemistry and thus avoid or minimize reequilibration impacts on corrosion scale and accumulated metals.

AFSs are classified as either portable or semipermanent. An example of each is provided in Figure 5-13. Portable devices are used on a temporary basis and connect directly to a 2½-in. fire hydrant nozzle. For these reasons, they are not usually a design consideration. Semipermanent devices are used in areas with known or anticipated chronic or seasonal water age or water quality control challenges. These devices utilize a cabinet-enclosed system mounted at grade. Installation and use of semipermanent AFSs require tapping the water main at the designated installation site and installing connective piping to the device inlet fitting. As part of the siting plan for these devices, utilities should consider ease of access, security, public safety, water drainage patterns, potential backflow issues, and provisions for electrical power and telemetry.

During system design, utilities should attempt to identify distribution system locations that may benefit from the use of semipermanent AFSs. Since these devices waste a fair amount of water to achieve their purpose, utilities should simultaneously explore other options to better manage water age or water quality that better conserve water resources.

Erosion corrosion. Managing residence time also involves ensuring that velocities going through mains are not too fast. Erosion corrosion has been observed in highly

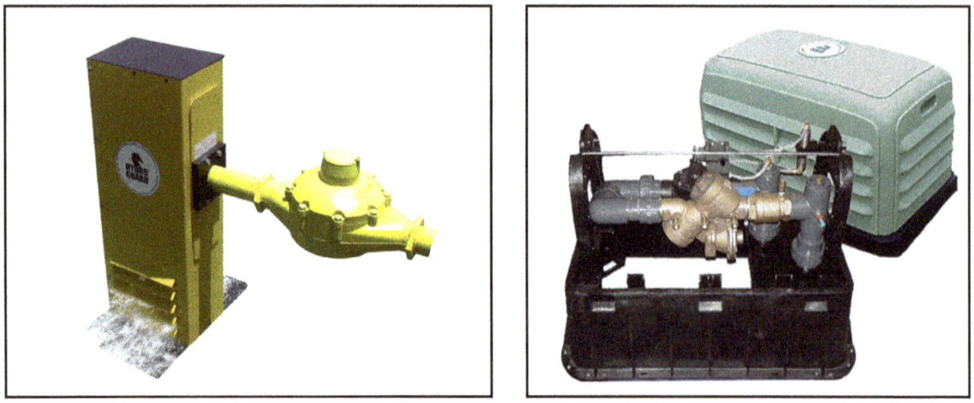

Courtesy of Mueller Co.

Figure 5-13 Portable automatic flushing station with 2½-in. flowmeter (left); semipermanent automatic flushing station (right)

tuberculated cast-iron pipes with high flow velocities or where abrupt changes in flow directions exist. The abrasive action of the fluid can scour the pipe scale and remove any protective covering that may have formed on the pipe wall. Water distribution systems designed to operate at lower flow rates will have reduced turbulence and, therefore, reduced erosion of the protective layer. However, care should be taken in the design of such systems to avoid very low flow rates, which promote excess material deposits and biofilm growth in stagnant areas.

Disinfectant booster stations. Maintenance of adequate disinfectant residual throughout the distribution system can be an important part of a corrosion control plan. Disinfectant residual helps to control biological activity, which is closely linked to microbiologically influenced corrosion (MIC), and plays a role in the oxidation–reduction and pH conditions within the water and at the pipe wall (Bremer et al. 2001, Friedman and Blain 2014, Schock and Giani 2005, Renner 2006, Reiber 2004). Although it is preferable to remove as much chlorine-demanding material from the water at the treatment plant to maintain disinfectant residual, disinfectant booster stations are sometimes required. These stations should be strategically located to influence the largest possible area of low disinfectant residuals and are often sited at the entry point to a pressure zone. In designing a disinfectant booster station, it is important to understand the hydraulics in the distribution system under a variety of operating conditions so that dosing can be carefully controlled and matched to system needs. Furthermore, it is important to install monitoring equipment, ideally with remote reporting capabilities, to have an understanding of influent disinfectant concentrations for dosing adjustment.

External Corrosion

This manual focuses on internal corrosion of pipes; however, it is briefly worth mentioning that external corrosion of mains and storage facilities are just as important.

Unprotected metallic pipes, such as iron or steel, that are placed in corrosive soils may deteriorate at a rapid rate. The outside wall structure will eventually weaken causing breaks to occur. Additionally, exterior corrosion will also increase when pipes are placed in soils where stray current may be abundant, such as in industrial or highly urbanized areas.

Microbial influence can also increase external corrosion. Conditions where soils or materials contain sulfate can influence the growth of sulfur-reducing bacteria, which will

Figure 5-14 Example of polyethylene casing for ductile-iron pipe

grow in anaerobic conditions. The microbial activity and by-products will result in an increase in external corrosion along the outside pipe wall (AWWA 2013).

To minimize corrosion, some utilities have started to replace metallic mains with PVC pipe. However, PVC is limited to the amount of internal pressure it can withstand and the maximum diameter, and it has a shorter expected life under normal conditions compared with ductile iron.

Loose polyethylene casings are commonly used along the outside wall of ductile-iron pipe (Figure 5-14). However, care must be taken during installation to prevent damage to the encasement, and all damage should be repaired to prevent direct pipe-to-soil contact. Polyethylene encasement material requirements are outlined in ANSI/AWWA C105/A21.5, which includes specifications for strength, tear resistance, impact resistance, and dielectric strength (AWWA 2010a). The specially manufactured polyethylene film is specified to be made from virgin resins and not contain recycled materials, as these may result in reduced film properties and also serve as a food source for bacteria involved in MIC.

External coatings, such as polyurethane, are a successful protective treatment for exterior corrosion protection. Along the same lines, cold-applied laminated tape and epoxy have been used to protect valves and other fittings of the distribution system.

For more information related to external corrosion control, refer to AWWA Manual M27 (2013).

Cathodic protection. Cathodic protection has also been used with success. Cathodic protection involves adding a sacrificial metal, in small quantities, near the pipe that will consume the stray currents faster than the ductile-iron material, thus diverting the corrosion potential to the sacrificial metal. Sacrificial metals or "anodes" are usually made of zinc or magnesium (Figure 5-15).

SYSTEM OPERATIONS AND MAINTENANCE

Routine system maintenance can reduce corrosion and/or remove the by-products of corrosion. Maintaining adequate disinfection residuals and routine cleaning of mains are key maintenance items that are described in more detail below.

Maintaining Disinfectant Residual Levels

Disinfectant residual concentrations decay as water passes through the distribution system because of reactions with materials present in the bulk water and at the pipe wall.

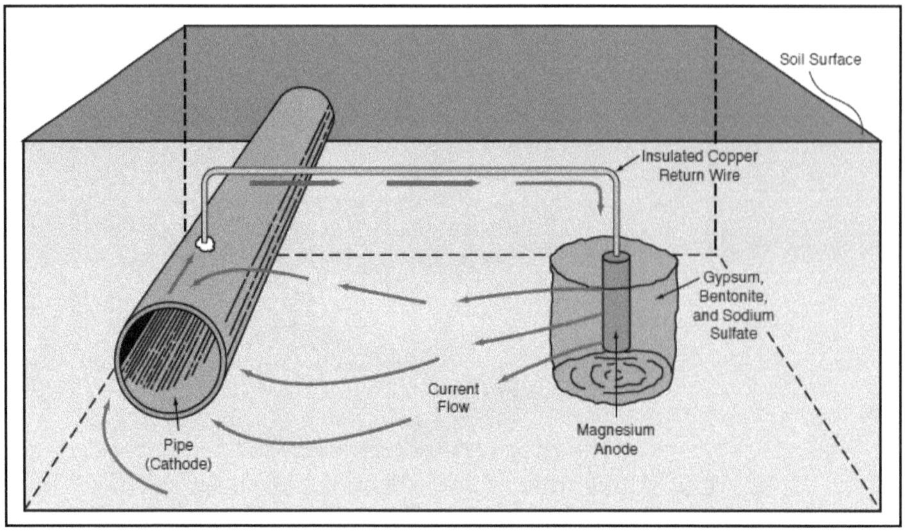

Source: *AWWA 2013, courtesy of NACE International*

Figure 5-15 Example of cathodic protection, using a sacrificial anode to protect the outer wall of a metallic pipe

The degree of residual degradation depends on the quantities of material available to react with the disinfectant (in the bulk water and at the pipe wall), temperature, and the time available for reactions to take place. The reaction time, or hydraulic retention time, was previously discussed in the "Managing Hydraulic Retention Time" section. The water temperature is largely outside the control of the water utility, although design of the system to avoid excessive heating of the water should be considered. This section discusses the type of operations and maintenance activities that can be done to increase the longevity of disinfectant residual.

At the treatment plant, long-lasting disinfectant residual can be promoted by removing chlorine-demanding material, particularly natural organic matter. Good control of pH and alkalinity can also result in low bulk decay for disinfectant residual. Once the finished water has left the treatment plant, decay is inevitable. Disinfectant-demanding materials present in the pipes include corrosion products, biofilm materials, and other contaminants that have intruded into the distribution system. In a synergistic way, cleaning out the accumulation of corrosion materials and biofilms will help to maintain disinfectant residuals, which in turn will help to manage corrosion and biofilm growth.

Main Cleaning

Main cleaning to remove legacy deposits from system piping is a distribution system maintenance practice and an industry-recognized best management practice. The most widely recognized benefits are water quality based and include enhanced disinfectant residual stability, reduced bacterial levels, and improved aesthetic water quality as reflected by customer reaction. These outcomes are typically cited by utilities as the primary drivers for conducting main cleaning. However, recent research focused on pipe–water interactions, both within the distribution system and in premise plumbing systems, have shed additional light on the wide-ranging impacts posed by legacy deposits and have demonstrated that effective routine main cleaning also plays a vital role in controlling internal corrosion, limiting metals release, and protecting the integrity of system infrastructure (Hill et al. 2014, Schock et al. 2014).

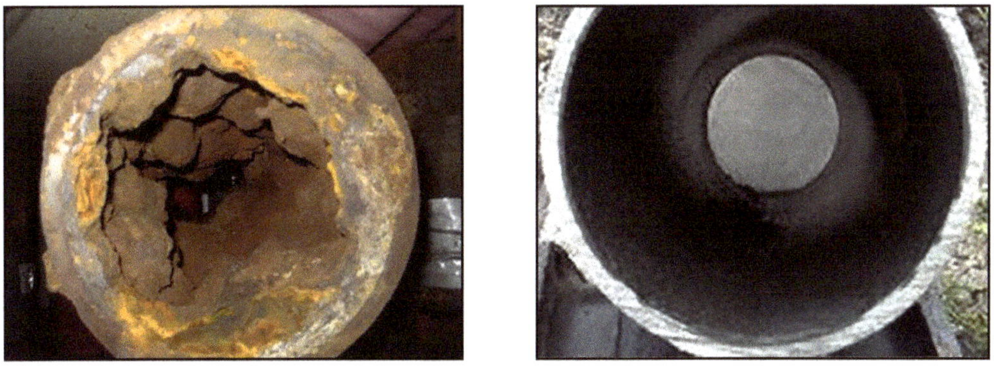

Source: *Friedman et al. 2010* Source: *Friedman et al. 2015*

Figure 5-16 Corrosion scale on unlined cast-iron pipe (left); metals-rich cohesive sludge on cement-mortar-lined ductile-iron pipe (right)

Impacts of legacy deposits on corrosion control. To understand the value of main cleaning as a corrosion control measure and identify an appropriate cleaning strategy for a specific application, it is important to consider the nature of legacy deposits and impacts posed by their presence on pipe surfaces. Deposits represent a mixture of accumulated solids of different origins, typically occurring in a complex, heterogeneous matrix. Component solids can include sediment, metal precipitates, biofilm, calcium scales, and various by-products of metallic pipe corrosion. The mass inventory, mineralogical composition, and mobility of deposits in a given system depend on numerous site-specific factors, including current and historical water quality; pipe type, lining, age, and condition; routine hydraulic conditions; maintenance history; and other factors (Friedman et al. 2003, USEPA 2006). For unlined metallic pipe, particularly cast iron (Figure 5-16, left photo), corrosion by-products comprise the majority of deposit mass, which is best quantified in tons per pipe-mile (Friedman et al. 2010). Non-scale-forming cement-mortar-lined pipe (Figure 5-16, right photo) and nonmetallic varieties of pipe (PVC, concrete, asbestos–cement, etc.) typically have much lower deposit mass inventories; however, because they lack the "dilution effect" produced by voluminous corrosion scales, accumulated inorganic solids on these mains can be substantially more concentrated with coprecipitated trace metals like arsenic and lead (Friedman et al. 2016, Hill et al. 2010, Lytle et al. 2004). Furthermore, because these solids originate from bulk water precipitation and deposition processes, they are often more vulnerable to physical and chemical mobilization (and associated water quality upsets) in comparison to adherent corrosion scale.

Although not all deposits are harmful per se, the only deposits recognized to be beneficial are the protective passivation layers and films produced on metallic pipes via corrosion, oxidation–reduction, and precipitation processes. These films limit the extent of corrosion and control the solubility of specific metals. Outer corrosion scale layers exist in a dynamic equilibrium with bulk water chemistry, such that stabilization of chemistry within certain ranges is critical to maintain scale stability. Chemistry stabilization is also important to prevent the soluble release of co-occurring metals via dissolution or surface desorption mechanisms. See chapters 3 and 4 for a more detailed discussion of water quality parameters impacting scale formation. While corrosion control treatment strategies seek to modify and control finished water chemistry at the system entry points, legacy deposits within the distribution system interact with bulk water through a variety of mechanisms (see Table 5-7) that can adversely impact these parameters and contribute to spatial and temporal water chemistry variations. These variations can destabilize scale and contribute to water quality deterioration (Hill et al. 2010). Therefore, the use of main cleaning to remove legacy deposits (while preserving passivation films); reduce microbial,

Table 5-7 Impacts of legacy deposits on water quality and corrosion control

- Deposits exert a disinfectant demand that lowers the residual concentration and oxidation–reduction potential (ORP). Low ORP can promote destabilization or dissolution of metal precipitates, corrosion scale, and passivation films, which in turn can result in metals release, discolored water, and accelerated re-corrosion of the exposed underlying pipe.
- Deposits, in particular corrosion scales and tubercles on cast-iron pipe, provide a large, protected environment in which bacteria can thrive. The tendency of deposits to consume disinfectant residual at the scale surface produces a disinfectant-free zone within the underlying scale. Bacteria that colonize the underlying scales include total coliform, nitrifiers, iron and sulfur bacteria, and certain pathogens. Microbes can produce acidic metabolites that increase local corrosion rates, impair scale integrity, consume alkalinity, and depress pH. Deposit and scale accumulation are major risk factors for nitrification in chloraminated systems.
- Deposits can concentrate various substrates and nutrients (ferrous iron, assimilable organic carbon, inorganic nitrogen, phosphate, etc.) that stimulate microbial activity and biofilm growth.
- Corrosion scales and inorganic precipitates (particularly those comprised of iron, manganese, aluminum, and phosphate) serve as substrates onto which various regulated trace metal and radiological contaminants (arsenic, lead, chromium, uranium, etc.) can concentrate via sorption/coprecipitation processes and later be released back into the bulk water at elevated levels (Hill et al. 2010). Lead levels up to 0.7% (dry weight basis) have been measured in deposits collected from distribution systems despite source lead levels <0.05 μg/L (Friedman et al. 2010).
- Deposits can coat metallic pipe surfaces and interfere with the formation of protective passivation films. For example, Schock et al. (2014) determined that deposition of manganese dioxide (MnO_2) solids on lead service lines obstructed formation of a protective plattnerite (ß-PbO_2) scale layer in one midwestern utility.
- Deposits and associated contaminants can be resuspended or released into the water column because of hydraulic, physical, and chemistry disturbances, causing potentially undetected elevated metals and/or bacterial counts, as well as discolored-water, taste, and/or odor issues.

Courtesy of Confluence Engineering Group, LLC

nutrient, and metal inventories; and maintain better uniformity of water chemistry plays a critical role in the overall framework of corrosion control.

Given the impacts listed in Table 5-7, utilities should evaluate system conditions and cleanliness as an integral part of a corrosion control assessment. In certain instances, erratic or elevated metals levels at customer taps can occur primarily as a result of inadequate system maintenance. Legacy deposits can accumulate substantial inventories of both unregulated and regulated trace metals over time. If these deposits are not routinely and properly removed, metals can be intermittently released in particulate or soluble form as a result of hydraulic and chemical variations typical of water system operation. In such cases, implementation of systematic flushing or cleaning for distribution system mains and/or premise plumbing may be able to control the problem without the need for treatment modifications. In other instances, main cleaning may complement a corrosion control treatment program to achieve more effective results and reduce risks of chemistry-based releases when new or modified water treatment processes are brought online to control corrosion.

Overview of main cleaning methods. Several main cleaning options are available to water utilities. The various choices can be categorized into two groups, as shown below and in Table 5-8. The options presented herein are not intended to be exhaustive but rather represent those cleaning strategies that are more commonly used.

- Group 1: stand-alone cleaning strategies.
- Group 2: cleaning strategies used specifically as a preparation step for pipe renewal (i.e., relining or rehabilitation).

The stand-alone cleaning methods in Group 1 vary in aggressiveness but are generally appropriate for most pipe materials and linings. When used on unlined metallic pipe, these methods should be applied in a manner that seeks to avoid exposing unstable scale

Table 5-8 Classification of main cleaning methods

Group 1: Stand-Alone Cleaning	Group 2: Cleaning as Preparation Step
Unidirectional flushing	Abrasive pigging
Air scouring	Mechanical scraping
Ice pigging	Hydraulic jetting
Foam swabbing	Power boring

material or bare metal. In contrast, when a utility plans to reline or rehabilitate metallic pipe, the highly aggressive cleaning methods in Group 2 are typically used to achieve complete removal of corrosion scale and produce a smooth, bare pipe surface. Highly aggressive methods are only suitable for structurally adequate unlined metallic pipe and must be immediately followed with relining or rehabilitation because exposure of the bare metal pipe surface may otherwise result in accelerated re-corrosion, disinfectant depletion, and extended periods of discolored water.

Conventional flushing has been purposefully excluded from Table 5-8, because this uncontrolled process is not recommended for main cleaning. In contrast to the highly organized process of unidirectional flushing, conventional flushing involves the practice of opening and flowing select fire hydrants without deliberate valve sequencing. As a result, flush water may be pulled from multiple directions and pipe segments at varying rates, resulting in a range of unknown velocities. It has the potential to create or exacerbate water quality problems by stirring up deposits in nearby portions of the system without removing them. Conventional flushing is intended for local bulk water turnover, typically as a quick-response measure to observations or complaints of poor local water quality, or to address stagnant flow conditions in dead ends and low-demand areas. It should be applied at a sufficiently low rate to ensure that velocities do not disturb or entrain deposits.

The following subsections provide additional details about each of the stand-alone cleaning methods (Group 1 methods) as well as abrasive pigging and mechanical scraping. Additional information on implementation mechanics and operational requirements can be found in AWWA's Manual of Water Supply Practice, Manual M28 (AWWA 2014c).

Flushing. Flushing is by far the most commonly used cleaning technique in the drinking water industry. Based on a 2002 survey of 500 utilities, 98 percent of respondents reported using flushing to clean distribution system mains (Ellison et al. 2003). Its appeal can be attributed to a variety of factors, including relative simplicity and low cost, effectiveness at addressing a variety of regulatory and customer-based water quality issues, and the ability to be performed without specialized equipment or system needs.

The two types of flushing are conventional (previously described) and unidirectional (UDF). UDF is a highly organized process that involves systematic planning and design of flushing routes for the purpose of controlled hydraulic scouring. Flushing routes consist of discrete stretches of main (typically between 1,000 and 2,000 lineal feet [305 and 610 lineal meters]) with defined end points and valve closures to ensure a unidirectional cleaning path. Multiple routes are used in series to enable a controlled cleaning progression from an initial clean water point throughout the targeted portion of the distribution system. Of the two flushing techniques, only UDF is intended for main cleaning. However, many utilities default to conventional flushing for all of their system flushing needs. For example, according to the same 2002 survey, 76 percent of respondents reported the use of conventional flushing for the purpose of cleaning mains versus just 22 percent that used UDF for that purpose (Ellison et al. 2003). This reflects a broad misunderstanding of the application roles for each method.

UDF is most effectively used as a periodic preventive maintenance practice to limit deposit accumulation and remove readily mobile solids. UDF is highly effective on loosely

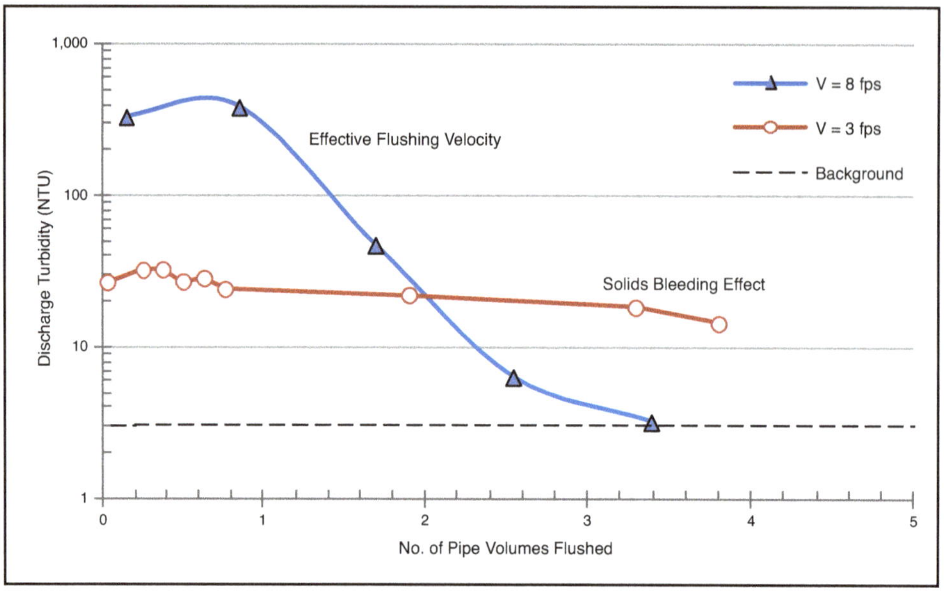

Courtesy of Confluence Engineering Group, LLC

Figure 5-17 Step-velocity test illustrating solids bleeding effect at suboptimal velocity

settled particles and sediment. Its effectiveness is generally limited on cohesive deposits such as biofilm, slime, and manganese film, although partial removal can be obtained under certain conditions. Flushing velocity is a critical, controllable variable that affects hydraulic boundary shear, the force responsible for scouring and lifting surface deposits. The optimal flushing velocity is application specific and involves striking a balance between maximizing the removal of targeted contaminants while protecting the pipe and any protective corrosion scale. Utilities can determine the optimal flushing velocity for a particular cleaning application by implementing a step-velocity test. In this test, two or more flushing velocities are applied in sequence on a defined stretch of pipe and accompanied with discharge water quality monitoring to assess incremental contaminant removal and pipe response.

The use of suboptimal flushing velocities can present unintended consequences. Velocities that are too low can result in impaired cleaning performance and cause a "solids bleeding effect" (see Figure 5-17). When this occurs, fractions of the native deposit are slowly mobilized over an extended period of time, resulting in significant water use and challenges restoring the clarity of the water. Conversely, excessively high flushing velocities can damage pipe or break tubercles from unlined iron pipe scale, potentially contributing to scale destabilization and discolored-water problems. In the case of heavily scaled iron pipe, it is important to account for average scale thickness and effective diameter when setting and calculating flushing velocity.

Air scouring. Air scouring is a patented cleaning technology that uses air along with low-rate flushing. In this process, compressed air that has been cooled, dried, and filtered is injected into an upstream hydrant in conjunction with low-velocity UDF, resulting in interspersed slugs of air and water (typically in a 75-percent-to-25-percent volumetric ratio) within the designated pipe segment. The viscosity difference between air and water produces a fluid-pulsing effect that enhances flow turbulence and scouring capability relative to flushing alone. Air scouring is appropriate for removal of cohesive deposits not otherwise amenable to removal by flushing alone, and/or if local discharge issues preclude the use of high-rate flushing. It may not be appropriate on weakened or corroded mains due to risks of main breaks and/or tubercle removal. The practice is typically provided by a service vendor that supplies and operates the air delivery system (compressor, cooler, filter). The utility provides the flushing-related equipment. The process typically involves service isolation to avoid introduction of air into home plumbing systems.

Ice pigging. Ice pigging is a patented, vendor-provided cleaning technology for water and wastewater mains. Developed in the United Kingdom in 2000, this technology was introduced to the North American market in 2012. The process involves injecting an ice slurry into a designated section of main (typically filling 20 percent of the designated pipe volume) through a hydrant to produce the ice pig. Using upstream system pressure, the ice pig is pushed through the target main to a downstream exit hydrant. The ice pig scours and gradually entrains accumulated deposits as it passes the pipe surface. The cleaning mechanism mimics that of a moving glacier and results in shear forces up to 1,000 times greater than use of water alone, thus allowing it to remove soft, cohesive deposits, biofilm, and friable scale. The ice slurry is preformed using utility water and NSF-grade salt (to depress the freezing point). The process produces a discharge residual stream that is concentrated in both suspended and dissolved solids and may require special disposal measures and/or permits.

A service vendor provides and operates a complete truck-mounted system that includes specialized equipment required to conduct the operation, including ice-making units, ice-holding and delivery units, process instrumentation/analyzers, hoses, diffusers, and appurtenances. The utility typically provides equipment associated with discharge residuals handling and disposal. Current ice delivery capacities are 2-ton, 10-ton, and 25-ton (Friedman et al. 2012).

Ice pigging is a particularly attractive alternative to flushing when dealing with tuberculated, unlined iron pipe. The City of Longview, Wash., recently implemented ice pigging to remove destabilized deposits from several miles of old cast-iron pipe following a source switchover. Previous attempts at UDF in the area had exacerbated metals release and discolored-water episodes by breaking protective tubercles and accelerating scale dissolution (see Figure 5.18, left image). The ice pigging operation effectively cleaned around the tubercles without damaging them, while removing 177 pounds per pipe-mile (50 kg per pipe-kilometer) of legacy deposits (Friedman et al. 2014). The preservation of tubercles was demonstrated by visual observation of pipe specimens that were removed before and after the ice pigging process, as shown in Figure 5-18.

Foam swabbing. Foam swabbing (also known as poly-pigging) involves the use of bullet- or cube-shaped devices made of soft, compressible bare polyurethane foam (see Figure 5-19). Low-density foam (1.5–2.0 lb/ft^3 [24–32 kg/m^3]) is used for lighter cleaning projects and to prove out the effective diameter of heavily corroded mains, while medium-density foam (5–8 lb/ft^3 [80–128 kg/m^3]) is used when slime, sludges, and/or friable scale are targeted for removal. In the process, one or more swabs are manually launched into an isolated main through a disassembled hydrant or dedicated pipe access station (refer to the "Accommodating Flushing and Cleaning" section earlier in this chapter for details on these stations) and propelled using upstream system pressure. Cleaning is achieved by the plowing action and frictional drag resulting from expansion of the swab against the pipe wall (AWWA Manual M28 [AWWA 2014c]). The swabs and dislodged solids are discharged through a downstream hydrant (see Figure 5-11) and may require special handling and disposal measures because of the large solids load. Multiple swab runs may be required depending on the nature and degree of buildup. Because of the concentrated slug of solids produced, utilities often elect to isolate customer services during the operation, although it is generally not required.

Abrasive pigging. Abrasive pigging is operationally similar to swabbing; however, because its purpose is to remove all solids (including scale) from unlined iron pipes, it involves a rigid foam bullet coated with a hard plastic shell and abrasive surface features

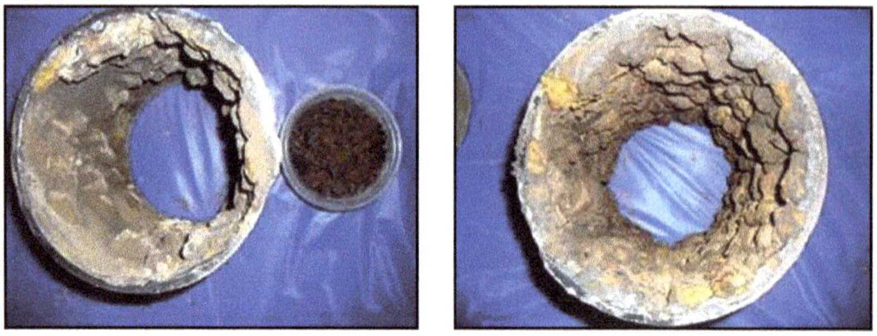

Courtesy of City of Longview, Wash.

Figure 5-18 Sections of unlined cast-iron pipe after UDF (left) and after ice pigging (right) indicate preservation of corrosion tubercles throughout the ice pigging process, whereas UDF caused exposure to weaker tubercles and pipe scales that led to increased discolored-water complaints

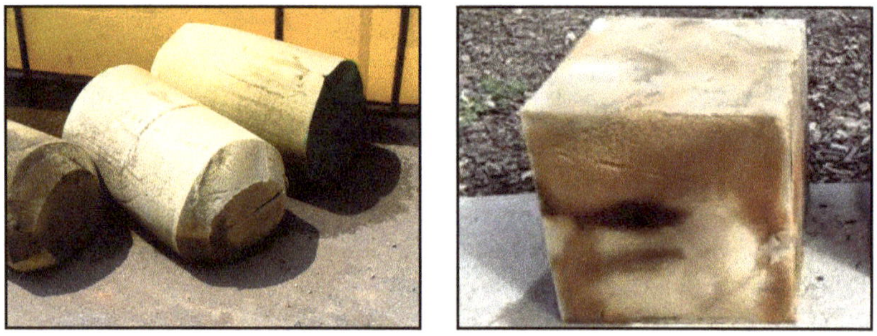

Source: *Friedman et al. 2016*

Figure 5-19 Bullet-shaped (left) and cube-shaped (right) polyurethane foam swabs

such as wire brushes, scrapers, or blades. These features provide highly aggressive forces to dislodge adherent scale. Depending on the internal condition of the main, multiple pig runs may be needed to produce a bare pipe surface suitable for relining or rehabilitation (AWWA Manual M28 [AWWA 2014c]).

The rigid nature of pigs requires installation and use of dedicated pipe access stations for pig launch and retrieval. During the cleaning and subsequent pipe renewal process, customer services are isolated and a provision for temporary bypass water supply should be provided. This process is most commonly used for deteriorated, corroded large-diameter unlined iron mains (12- to 36-in. diameter) where pipe renewal or rehabilitation is deemed more feasible or cost-attractive than replacement. However, if the pipe condition is severely compromised, pigging may not be appropriate unless structural rehabilitation is planned. Abrasive pigging should not be performed for plastic or cement-lined pipe. The process is typically contracted out because of the specialized nature of the operation and equipment needs.

Mechanical scraping. Mechanical cleaning of water mains can take one of several forms, but the basic concept involves use of a mechanically driven metal device pushed or pulled through a designated pipe segment. As with abrasive pigging, its purpose is to scrape and remove all deposits and scale to expose bare metal pipe for follow-up relining or rehabilitation. A typical device consists of a steel frame shaped like a piston attached to a drag cable (see Figure 5-20). Some devices incorporate tempered steel blades to increase aggressiveness. Multiple device passes are used as needed to achieve total scale removal (AWWA Manual M28 [AWWA 2014c], Friedman et al. 2012).

Courtesy of South Central Connecticut Regional Water Authority

Figure 5-20 Mechanical scraping head used to clean pipe scale

The operation is carried out with the main dewatered. Pipe access pits are installed (typically every 500 ft [152 m] and at junctions) to facilitate installation of the cable and scrapers. During the cleaning and subsequent pipe renewal process, customer services are isolated and a provision for temporary bypass water supply is typically provided due to the extended nature of the operation. This process is most commonly used for deteriorated large mains (12- to 36-in. diameter) where pipe renewal is deemed more feasible or cost-attractive than replacement. The process is typically contracted out because of the specialized nature of the operation and equipment needs.

Reducing Mobilization of Accumulated Metals by Controlling Shear Stress

It has been demonstrated that layers of cohesive inorganic-laden biofilm material form under prevailing hydraulic conditions within pipes, including in nonmetal pipe materials (Douterelo et al. 2013). The accumulated metals can represent many years of corrosion and transport of metals, and when mobilization of this material occurs, the metals released can be significant. This behavior has been observed in both metal pipes, where tuberculation is contributing directly to metal accumulation, and in nonmetal pipes that are the site of accumulation due to hydraulic conditions and the transport of metals from elsewhere in the distribution system.

When operating distribution systems to manage the impact of corrosion and material deposits, it is essential to consider the forces on the pipe wall where accumulation of metals occurs. That is to say, layers that remain on the pipe wall are those with a cohesive strength exceeding local shear stress forces. This effectively "conditions" the layers to a strength equivalent to historical shear stress. The shear stress can be elevated above that experienced in recent operation as a result of an increase in flow/velocity from a disruptive water system event such as a water main break or in a planned flushing program. With increased velocity, the cohesive strength of accumulated layers is no longer sufficient to hold them at the pipe wall, and consequently, material is eroded and mobilized into the bulk flow (Husband and Boxall 2010, van Thienen et al. 2011, Hossain et al. 2011).

Shear stress is not simply a function of water velocity. Shear stress is proportional to the head loss multiplied by the diameter in a given pipe and is expressed in units of force per area (e.g., newtons of force per square meter of pipe wall surface, as shown in Figure 5-21). For two pipes experiencing the same increase in flow velocity, the one with higher head loss (e.g., smaller diameter, greater roughness) will experience a greater shear stress increase. It is the *change* in shear stress, above the conditioned state, that results in mobilization of material.

Pipes that experience a diurnal variation in flow have less buildup of biofilm and particulate material than constant-flow pipes, even at equivalent average shear stresses. This is due to regular mobilization of accumulated materials (Douterelo et al. 2013).

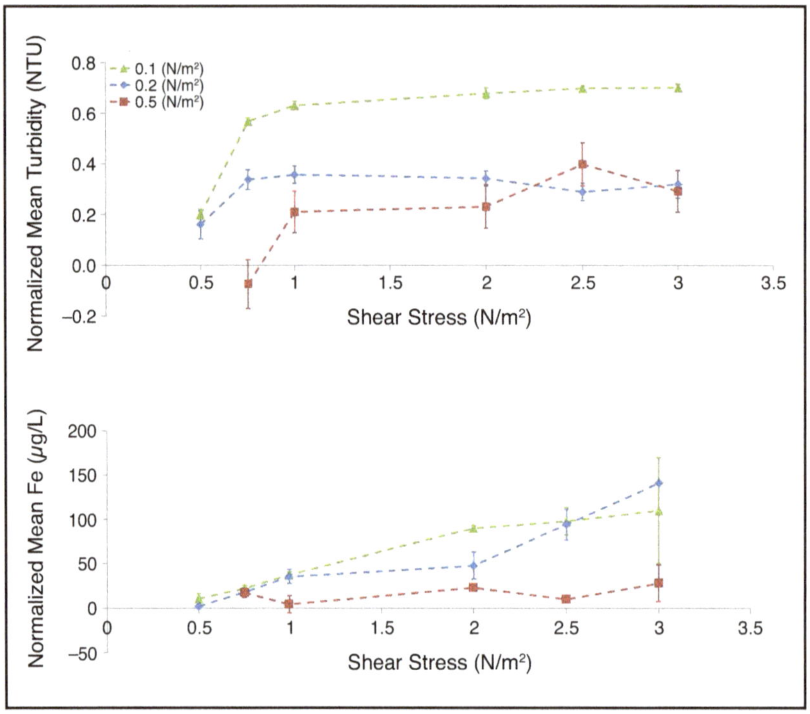

Note: 1 newton of force = 0.22 lb of force.

Source: *Speight et al. 2013*

Figure 5-21 Results for material mobilized in terms of turbidity and iron (y-axis) versus shear stress of flushing (x-axis) following conditioning at three shear stresses

The concept is illustrated in Figure 5-21, which shows that the turbidity and iron concentration of material mobilized from an experimental pipe loop during flushing is greatest when the shear stress during conditioning was lowest.

Although there is not a linear relationship of shear stress to flow rate (or velocity), measuring flow is an acceptable surrogate for shear stress if the diameter of the pipe under evaluation remains constant. Similarly, turbidity measurement acts as an accurate surrogate for metal concentrations, particularly iron (Cook et al. 2015). With a record of the flow and turbidity, it is possible to explain and, more importantly, to predict the distribution system behavior. Figure 5-22 shows a year's flow and turbidity from a 9.3-mi (15-km), 30-in. (800-mm) steel transmission main. During the first half of the year, the flow regularly reaches 20 million liters per day (ML/d), which is approximately 5.3 million gallons per day (gpd), such that the corrosion and deposits are conditioned to this value. No increases in turbidity are seen when flows remain at or below the 20-ML/d level. However, during the summer months, the flow reduces and material accumulation occurs. It can be observed that a number of short-term elevated flows during the summer produce turbidity responses (seen as spikes) because accumulating material that is now conditioned to the lower summer flows is being stripped from the pipe walls. The greatest turbidity response, however, occurs when the flow increases in October to exceed the 20-ML/d value that had previously conditioned the accumulation of material.

The example from Figure 5-22 illustrates the potential for management of material accumulation and its release to customers by controlling shear stress. If the flow can be deliberately controlled, not only can potential incidents be avoided, but by using carefully managed incremental flow increases, the accumulated material can also be gradually

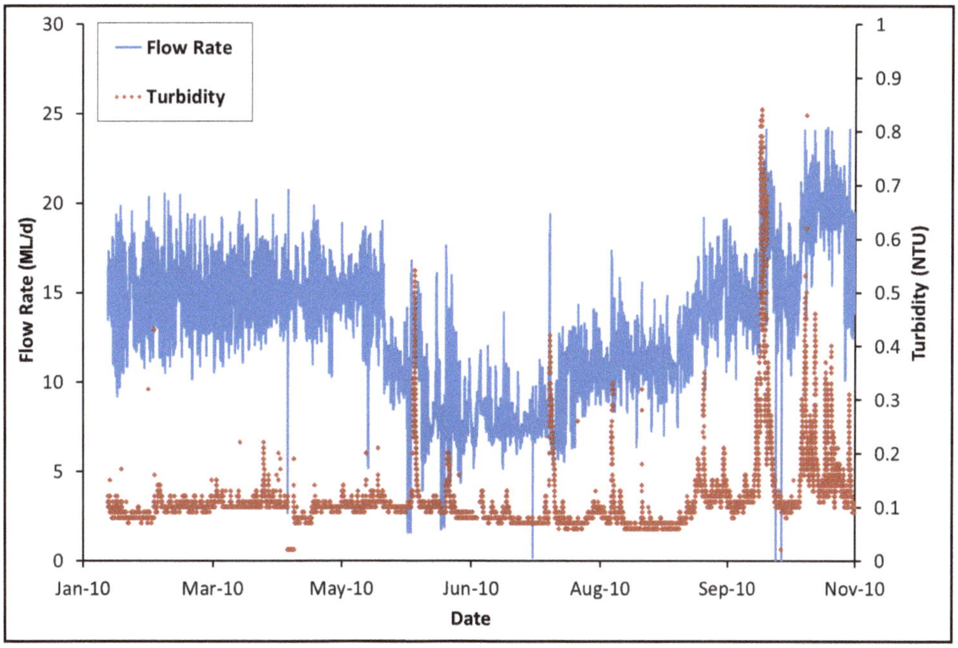

Figure 5-22 Flow and turbidity record for a transmission main segment

eroded during normal operation without any impact to customers. This practice is analogous to flushing but without discharging the water. The required flow rate for slowly eroding accumulation will depend on the flow history and accumulation rate within the pipe.

The implementation of controlled shear stress increases must be evaluated on a case-by-case basis to ensure that (1) the selected pipes are able to handle the increased flows, (2) flow control and measurement equipment is in place, and ideally (3) turbidity is monitored to prevent excess mobilization of metals. Not all material can be removed by this process, but a reduction in discolored-water complaints and sediment buildup in the pipes can be achieved.

PIPE REHABILITATION

Pipe Linings and Coatings

In addition to selecting the proper piping material to minimize corrosion, one technique to keep corrosive water away from the pipe wall is to line the wall with a protective lining or coating. With new pipe, linings and coatings are usually applied during the pipe manufacturing process, but they can also be applied in the field when the pipe is installed. Some linings and coatings can be applied even after the pipe is in service.

Linings can also be installed to provide structural integrity or improvement to a pipeline that has been in service for a length of time. Linings are categorized with structural classifications identified as Classes I through IV, with Class I linings being nonstructural (coatings and nonstructural linings) and Class IV linings being structurally independent. These classifications are further defined in AWWA Manual M28, *Rehabilitation of Water Mains*, Appendix A (AWWA 2014c). As an example, a Class I lining could be an epoxy, a cement–mortar lining, or a thin polyethylene terephthalate (PET) liner installed only to serve as a barrier between the water and the pipe, preventing interaction between the two.

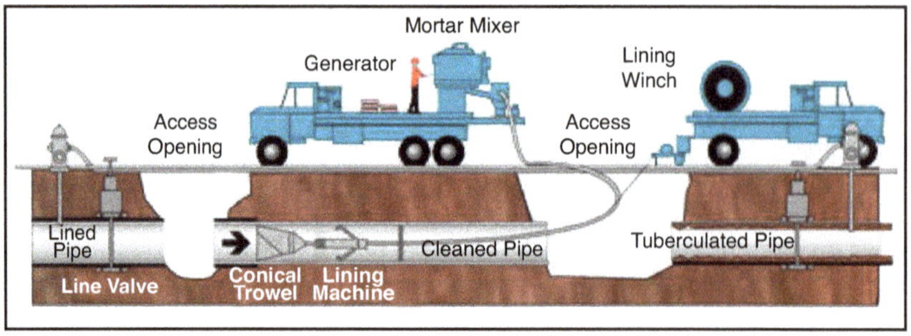

Figure 5-23 Application of cement–mortar lining

Figure 5-24 Example of pipe before and after cleaning and lining with cement mortar

A Class IV lining could be a cured-in-place pipe installed to replace a severely corroded pipe segment, which does not rely on the structural integrity of the host pipe to function.

The most common pipe lining materials are epoxy paint, cement mortar, and polyethylene. It is common practice to reline cleaned pipe to protect the newly exposed pipeline material. The most common technique is to apply a cement–mortar lining. Figure 5-23 illustrates one process to apply a cement–mortar lining. Spray-on epoxy linings are especially useful in waters that are aggressive to cement linings, typically high negative Langelier indexes. Figure 5-24 illustrates iron pipe that has been cleaned and relined.

Cement mortar has an expected life of 50 years. Epoxy linings are estimated to last between 40 and 60 years (Deb et al. 2006) but are much more expensive to apply. Costs will vary by geographical location, as some materials are more readily available in certain areas. The effectiveness of the coatings is also limited by the degree of pipe cleaning and smoothness of the pipe after cleaning. Remaining tubercles after cleaning can cause inconsistency in the pipe coating. In addition, when lining pipes, the strength of the pipe has to be able to withstand the cleaning process. Effective coatings cannot be applied to ruptured pipes.

Figure 5-25 illustrates steel pipe with an epoxy lining. Table 5-9 summarizes the most commonly used pipe coatings and their recommended applications.

Main Replacement

A utility may determine that pipe rehabilitation techniques are not viable in specific circumstances. If a system has trouble with highly aggressive water or soils in contact with the outside of a main or an established history of corrosion damage in a specific distribution zone inside the main, the decision may be made to replace pipe segments as a result of damage caused by corrosion. There are certainly economic costs associated with this work, specific to the materials and labor involved with installation of a new pipe in the

Source: *Hock et al. 1999*

Figure 5-25 Steel pipe with epoxy coating

ground. It is also important to account for those noneconomic costs related to the work, specifically those societal (e.g., disruption to traffic) and environmental (e.g., waste created from tearing up asphalt roads) costs that can sometimes be difficult to quantify. Once a utility is comfortable with these costs and has made the decision to replace pipe in the system, the decision on the pipe material to be used can be addressed. AWWA has several manuals of practice that are available to assist in the decision on what pipe material to install.

The decision to replace a water main segment is highly specific to a utility and the distribution system. Some areas may face high levels of customer resistance to open-trench pipe installation work. Other areas may find that interference from other nearby utilities make trenching and pipe installation difficult. These situations tend to suggest that rehabilitation of an existing pipe would be beneficial, but the final decision still rests with the individual company.

When replacing a main that is connected to an LSL, it is highly recommended to replace the full service line as well. Advanced customer notification is required, and customers should be strongly urged to replace the private portion of the LSL if not owned by the utility. The next two sections discuss more details related to LSL replacement and customer communication.

Linings and Coatings for Lead Service Lines

Most of the LSLs still in use in the United States and Canada are structurally sound and would be expensive to replace. Several utilities have considered the possibility of lining or coating LSLs to reduce or eliminate lead release, thereby protecting human health and, in some cases, achieving compliance with the Lead and Copper Rule or provincial guidelines. Some utilities have conducted pilot or demonstration tests in the hope of identifying a commercially available lining or coating technology that is not only effective but can be implemented at significant cost savings relative to the cost of replacement. To date there has been minimal use of such technologies in the United States and Canada, but greater use has been reported in the United Kingdom and several other countries. As technologies evolve and circumstances change, lining and coating of LSLs in the United States and Canada may become commonplace, especially for LSLs that cannot be readily replaced using conventional methods, such as trenching, because of their depth or location (e.g., proximity to other utility lines, historic buildings, or busy traffic ways).

Many challenges must be overcome and there are many factors to consider when deciding whether lining or coating LSLs is the best alternative in a given location. The most obvious challenge in lining or coating LSLs is their small diameter. Although some LSLs, especially those providing water to commercial establishments, have an inner diameter (ID) of 1 to 2 in., most residential LSLs have an actual ID in the range of 0.50 to 0.75 in. This precludes the use of many of the lining and coating technologies summarized in

Table 5-9 Pipe linings—Summary of applicable technology and recommended use

Technology	Recommended Application
Cement–mortar lining	• Prevents scale formation and internal corrosion; reduces pipe roughness and friction. • Considered with hydraulic and discolored-water/taste and odor problems when there are no structural and joint leaks and original pipe material is cast iron, ductile iron, or steel. • Should not be considered when soft or acidic water is conveyed because of possible deterioration of the cement–mortar lining.
Epoxy resin lining	• Protects original pipe against corrosion and reduces roughness and friction. • Considered with hydraulic and discolored-water/taste and odor problems when there are no structural and joint leaks.
Conventional slip lining	• Effective diameter of pipe is reduced, but with new pipe having a smoother interior surface. • Excavations are required for service connections, entrance pits, and exit pits. • Various pipe materials (ductile iron, PVC, HDPE, and steel) may be used as new pipe. No strength is added to the host pipe in conventional slip lining.
Close-fit slip lining	• Classified as structural or semi-structural lining depending on the thickness of the liner. The inserted pipe adds strength, prevents further internal corrosion, and reduces pipe roughness and friction. • Considered for hydraulic, joint leak, and discolored-water/taste and odor problems when no structural issues are involved.
Cured-in-place pipe	• Compared to close-fit lining, the thickness of cured liner is typically less than a close-fit liner. • As with the close-fit liner, the loss of diameter is compensated for by a reduction in roughness and friction. • As opposed to epoxy lining, cured pipe also provides a certain measure of leakage protection. • Considered a semi-structural liner and is applicable for hydraulic, joint leak, and discolored-water/taste and odor problems when no structural issues are involved.
Pipe bursting	• Pipe bursting is not a structural lining technique but should be mentioned because it is considered suitable for cast-iron, PVC, asbestos–cement, and thin-wall steel pipes. • Pipe bursting recommended for deep mains with sufficient cover to avoid heaving. • Host pipe should not have offset pipe joints or clamps with bolts. • Applicable for replacing pipes of the same diameter or larger. • Excavations are required for service connections, entrance pits, and exit pits.

Source: *USEPA 2002*

Table 5-9 and recommended by AWWA (2014c). In some cases, this is because the equipment manufacturer does not make equipment small enough for LSLs. For example, most manufacturers of spray-applied coatings do not make spray heads small enough to fit inside a pipe having an ID less than 3 to 4 in. In other cases, the manufacturer may choose not to make a technology available for such applications. For example, some types of plastic tubing are available in sizes that could be slip-lined into an LSL, but the manufacturer does not market them for this purpose because the potential market and profit margins are considered too small. In some cases, the lining or coating would be thick enough to significantly increase head loss (reduce flow) such that consumer satisfaction would suffer and complaints would increase. Some utilities specify a minimum ID of 0.75 to 1.0 in. for new service lines and are reluctant to adopt or approve practices inconsistent with their current standards.

Commercially available technologies potentially suitable for lining or coating small-diameter pipes, especially LSLs, were reviewed by Kirmeyer et al. (2000) and Randtke et al. (2014). The two technologies currently drawing the greatest interest in the United States and Canada, and in other countries having significant numbers of LSLs, are epoxy coatings and PET liners. Both technologies are commonly used in commercial and industrial settings where pipe replacement would be prohibitively expensive or disruptive (i.e., situations similar to those often encountered with LSLs). There is also growing interest in polyurethane- and polyurea-based coatings, due in large part to their relatively rapid curing times, and it would not be surprising to see such products effectively competing for the LSL market in the near future.

Other challenges facing decision makers considering lining or coating LSLs include the following:

1. **Technical and logistical challenges.** For epoxy coatings to properly adhere, the surface must be clean, dry, and roughened. This is normally accomplished by sandblasting with compressed air. Cleaning requirements are less stringent for PET liners, but pigging or scraping is often necessary. For both technologies, the presence of heavy deposits inside the LSLs, as may occur downstream from corroding unlined cast-iron mains, can make it impractical to adequately prepare LSLs to be lined or coated.

   Curing times for epoxy coatings can be relatively long, 12–16 hours according to AWWA (2014c). This can adversely impact the productivity of work crews (i.e., cost) and consumer satisfaction with the installation process. For this reason, some manufacturers have developed epoxies that cure more rapidly. PET liners are cured at the factory and are simply inflated inside an LSL using hot, pressurized water. Thus, same-day return-to-service installation is available for both PET liners and some epoxy coatings.

   For epoxy coatings, applying a uniform coating while avoiding excessive pooling and premature hardening (especially in cold weather) can be challenging. When installing liners, it can be difficult to negotiate sharp bends (i.e., bends with a radius less than the bending radius of the liners, which tend to be encountered near connection points) and sharp protrusions associated with welds or joints can tear the liner, resulting in installation failure. Experienced distribution system employees may be able to estimate the frequency with which such conditions are likely to be encountered.

2. **Economic challenges.** Achieving a reasonable economy of scale can be difficult. Lining or coating a single LSL, or even a single group of LSLs, is unlikely to be cost-effective relative to the cost of having an experienced crew install new service lines. The cost per connection will be lower if clusters of LSLs are lined or coated, especially if each cluster contains the number of connections a work crew can install in one day, and even more so if a systematic lining or coating program is implemented such that one or more trained crews can continue working day after day for an extended period of time.

   Reliable cost estimates are difficult to obtain. Although lining and coating materials are much less expensive than copper tubing (the material most commonly used for service lines), the cost of materials represents only a small fraction of the total cost of lining or coating an LSL. Other costs (for labor, street repair, traffic control, etc.) and conditions influencing costs (the length and burial depth of the LSLs, soil conditions, the location of other utility lines, etc.) will vary significantly from one location to another. Furthermore, some technologies are licensed to individual contractors, in which case the cost will depend on the location and availability of a licensed contractor, the willingness of a local contractor to apply for a license and purchase the necessary equipment, or the cost of bringing in an experienced installer from another location.

Lining or coating an LSL makes sense only if the cost of doing so is less than the cost of replacing it, and only if the service life of the lining or coating is sufficiently long (i.e., 25 to 50 years or more for a typical utility). Although PET liners and epoxy coatings are expected to last a very long time, most manufacturers warrant their products for much shorter periods of time. This puts utilities in the position of either negotiating an extended warranty (most likely at a higher cost) or accepting some risk that the service life of the product will be less than expected. Further complicating this situation is that the service life of most linings and coatings depends primarily on the quality of the installation and the properties of the underlying substrate.

Most utilities with LSLs recognize that partial LSL replacement can result in elevated lead levels at the tap for an extended period of time (USEPA 2011) and that full replacement would provide superior long-term public health protection. Some utilities, such as the city of Saskatoo, Saskatchewan, offer their customers some sort of cost-sharing incentive to encourage full replacement, while others (Washington, D.C.) amplify the health implications when LSLs are not fully replaced, but only a small percentage of homeowners remove their portion. Some utilities are evaluating LSL lining and coating alternatives in the hope of offering their customers a lower-cost and less-disruptive option for reducing or eliminating lead release from the customer-owned portions of their LSLs.

3. **Regulatory challenges.** A PET liner or epoxy coating can greatly reduce or eliminate release of lead from an LSL, but elevated lead levels may still exist if only the utility-owned portion of the LSL is lined or coated or if lead continues to be released from deposits that have accumulated in the interior plumbing. Thus, although LSL linings and coatings are expected to reduce lead release (and lead concentrations in the tap water) over time, they do not guarantee that lead exposure will be eliminated. In addition, current regulations do not explicitly address replacement of LSL linings and coatings.

   Some individuals have expressed concern that chemical constituents in a given lining or coating material, or their by-products, may leach into the water and cause regulatory or water quality issues (e.g., exceedances of current or future health-based levels or taste and odor problems). To a large extent, this concern is already being addressed by the widespread North American practice of requiring materials used in drinking water distribution and plumbing systems to comply with NSF/ANSI Standard 61 (NSF International 2015). This standard requires testing of materials to demonstrate that the concentrations of a broad spectrum of chemicals (including currently regulated chemicals as well as other chemicals of concern that may be regulated in the future) do not exceed minimum threshold levels.

   It is conceivable, of course, that new chemical contaminants associated with lining or coating materials may be identified in the future and action may be required to address them. However, PET liners and epoxy coatings are in widespread use in applications involving contact with food, beverages, and medical and pharmaceutical products. Consequently, in all likelihood, water utilities would not be alone in having to address such a problem. In addition, a lining or coating can effectively reduce leaching of other toxic chemicals or metals besides lead that have accumulated in pipe deposits (e.g., Lytle et al. 2004, Schock et al. 2008). For example, Breault (2014) found that although a PET liner leached trace amounts of antimony into standing water (an order of magnitude less than the maximum contaminant level for antimony), much higher levels of antimony were found in samples drawn from unlined (control) pipe sections. In a 12-week study comparing in situ epoxy coating to chemical treatment for the control of lead and copper corrosion, Hock et al. (1999) found very little leaching of organic contaminants from the epoxy-lined pipes as long as the epoxy was allowed to cure completely and the lines were flushed prior to first use.

Lead Service Line Replacement

Lead pipe was used for service lines in the United States and Canada during a period that began before 1900 and continued to the middle of the 20th century. Today, many LSLs still exist in older cities. A utility should replace the LSL in the following instances:

- Service line break
- Main replacement
- Exceed drinking water regulations for lead
- Remove LSLs from the system to reduce exposure of the building occupants to lead
- Other infrastructure work, such as sewer or gas main renewal that may cross LSLs

Partial lead service line replacement (PLSLR) is the replacement of the utility-owned portion of the service line while keeping the original customer side in place. PLSLRs are highly discouraged over the preferred method of a full lead service line replacement (FLSLR; utility and customer portions of the service line) but can occur when the homeowner refuses to have their service line replaced or the utility does not have the means to replace the customer's portion under certain circumstances such as emergency service line break repairs.

PLSLRs have the potential to increase lead release, especially from the release of particulate lead. During replacement of the LSL, the associated lead scale remains in the customer side of the service line and in the premise plumbing. These scales are disturbed after PLSLR and can become dislodged. Even FLSLRs can create vibration and other disturbances that could cause dislodgement and release of scale and other particulate matter inside the premise plumbing and appurtenances (Brown and Cornwell 2015, Del Toral et al. 2013).

Experience with LSL replacement, especially partial replacements, has indicated that it can take several months after the disturbance for the lead releases to be reduced to acceptable levels, chiefly because of sporadic releases of particulate lead (DC Water 2013, Providence Water 2013, USEPA 2011).

If an LSL or a portion of the line is to be replaced, several critical steps must be taken before, during, and after the process, as outlined in Brown et al. 2015:

1. **Notify the customer of the replacement.** Once it is determined that at least a portion of an LSL is going to be replaced, personal contact between utility staff and the customer should be initiated. Information provided should include description of the project (what you are doing, why, how long it will take, etc.) and any precautionary notes. This communication should include encouragement to customers to replace their portion of the LSL in conjunction with water system replacement of the utility portion of the LSL. Customer communication and notification of replacement should be done well in advance, and several times thereafter, to allow the customer sufficient time to arrange for their own portion of the LSL to be replaced. Examples of appropriate customer communication techniques

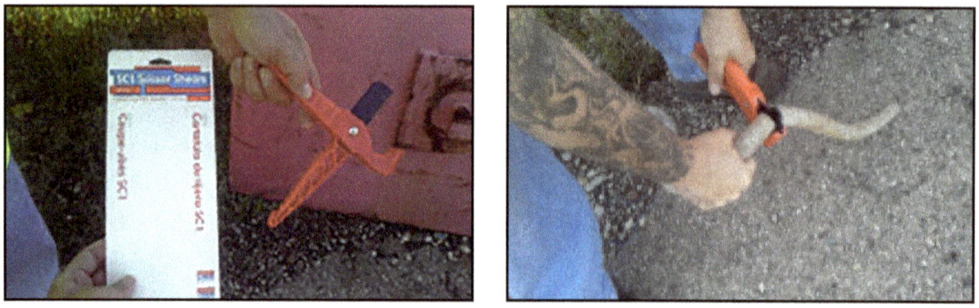

Figure 5-26 LSL cutting tool used to reduce cutting and disturbance during LSL replacement

and tools can be found in AWWA's *Communicating About Lead Service Lines: A Guide for Water Systems Addressing Service Line Repair and Replacement* (AWWA 2014a).

2. **Take precautions to minimize pipe vibration when replacing the lead service line.** If removing only a portion of the LSL, a cutting tool like the one shown in Figure 5-26 is preferred to other available tools that use a sawing motion or other abrasive action. If copper is used to connect to the remaining LSL on the customer side, it may be necessary to install a dielectric to protect against or reduce galvanic corrosion effects, though this will also mean taking precautions to limit potential hazards from electrical shock. Water service pipes are often used for bonding home electrical systems, and home wiring may be deliberately or inadvertently grounded to the water supply line. These connections to the service line can place an electric current on the service line (Welter et al. 2013). The current National Electric Code (NFPA 2011) provides for bonding to water service lines (Section 250.104) and requires at least 10 ft (3 m) of metal (water) pipe in direct contact with soil (Section 250.52(A)(1).

   AWWA policy opposes the grounding of electric systems to pipe systems conveying drinking water to a customer's premises (AWWA 2014b). Specifically, AWWA opposes the systematic interchange of stray electric current between electrical distribution systems and water distribution systems, as well as the use of the water pipe system as an essential part of any electrical system. AWWA does not object, however, to the connection of all interior piping of a building to the electrical service neutral and to a separate grounding electrode, provided that such interior pipe systems and grounding connections are electrically isolated from the water supply pipe system.

   When the water service line is serving as grounding for the home, actions that shorten the required 10 ft (3 m) of metal pipe in direct contact with soil are important to the structure's compliance with the current electric code.

   Precautions must also be taken to minimize pipe vibration, even during a full LSL replacement, because lead adsorbed onto indoor plumbing, such as galvanized pipes, and scales containing other adsorbent metals, has the potential to expose customers to higher lead concentrations (McFadden et al. 2011).

3. **Flush the lines.** Flushing following LSL replacement should always be provided. Research described in Brown and Cornwell (2015) suggests greater benefit from a longer duration and higher-velocity flushing procedure than often used after LSL replacement. This specifically includes high-velocity flushing of an outdoor tap for 10 minutes followed by 30 minutes of flushing for inside taps with all taps opened at the same time and with screens, aerators, filters, and so on, removed. It is also highly recommended for the homeowner to use a home water filter (the device should be certified as meeting NSF/ANSI Standard 53 for lead removal

[NSF International 2009]) for direct drinking water consumption such as drinking water, baby formula, cooking, and dental care (Brown and Cornwell 2015). Some products certified to Standard 53 may not be certified specifically for lead removal but for other contaminants. Products certified to Standard 53 for lead removal must explicitly identify lead as one of the contaminants removed.

4. **Collect samples after four to six months.** Post–LSL replacement, sampling four to six months after replacement may be a useful precaution for water systems to consider. Taking these samples will involve re-engaging the customer. Prior to conducting this sampling, the utility should determine an acceptable goal for lead levels in these samples and a plan of action if lead levels exceed this goal (i.e., provide water filters, resample at a later date, conduct a plumbing profile to determine source of elevated lead, etc.). Water systems should be prepared with recommended practices and precautions for customers in situations where lead levels are above utility goals in any of these samples.
5. **Record location.** Maintaining a record of the replacement and characterizing the service line will be beneficial for compliance with regulatory requirements, future interactions with customers at that address, and the system's asset management planning.

CONCLUSION

Distribution treatment techniques to minimize corrosion in the distribution system and premise plumbing can be accomplished by shortening hydraulic retention times, ensuring adequate secondary disinfectant residual, and maintaining proper flow. The most efficient way to achieve these goals starts with a good distribution design. Minimizing dead ends and installing storage tank mixers, automatic flushing units, and chlorine booster stations are primary considerations. In addition, development and implementation of an operations and maintenance program that includes routine unidirectional flushing, pigging, and pipe rehabilitation will minimize sediment buildup and tuberculation, control biofilm, and minimize water quality degradation from the treatment plant to the customer's tap.

ADDITIONAL READING

Communicating About Lead Service Lines: A Guide for Water Systems Addressing Service Line Repair and Replacement. American Water Works Association (AWWA), 2014.

Contribution of Service Line and Plumbing Fixtures to Lead and Copper Rule Compliance Issues. Water Research Foundation (WRF), 2004.

Controlling Lead in Drinking Water. WRF 4409, 2015.

External Corrosion Control for Infrastructure Sustainability, Manual M27, 3rd ed. AWWA, 2014.

Galvanic Corrosion Following Lead Service Line Replacement. WRF 4349, 2013.

Managing Change and Unintended Consequences: Lead and Copper Rule Corrosion Control Treatment. AWWA, 2005.

Rehabilitation of Water Mains, Manual M28, 3rd ed. AWWA, 2014.

REFERENCES

AWWA (American Water Works Association). 2010a. C105/A21.5. *American Standard for Polyethylene Encasement for Ductile-Iron Pipe Systems*. Denver, Colo.: AWWA.

AWWA. 2010b. *Water Transmission and Distribution: Principles and Practices of Water Supply Operations*, 4th ed. Denver, Colo.: AWWA. pp. 10, 23, 368.

AWWA. 2013. Manual M27. *External Corrosion Control for Infrastructure Sustainability*, 3rd ed. Denver, Colo.: AWWA.

AWWA. 2014a. *Communicating About Lead Service Lines: A Guide for Water Systems Addressing Service Line Repair and Replacement*. Denver, Colo.: AWWA.

AWWA. 2014b. *Grounding of Electrical Circuits on Water Pipe:* Policy. Denver, Colo.: AWWA.

AWWA. 2014c. Manual M28. *Rehabilitation of Water Mains*, 3rd ed. Denver, Colo.: AWWA.

AwwaRF and TZW (Awwa Research Foundation and DVGW-Technologiezentrum Wasser). 1996. *Internal Corrosion of Water Distribution Systems*, 2nd ed. Denver, Colo.: AwwaRF.

AwwaRF and USEPA (Awwa Research Foundation and US Environmental Protection Agency). 2008. *Contribution of Service Line and Plumbing Fixtures to Lead and Copper Rule Compliance Issues*. Denver, Colo.: AwwaRF.

Berend, K., and T. Trouwborst. 1999. Cement–mortar pipes as a source of aluminum. *Journal AWWA*, 91(7):99.

Breault, Z.A. 2014. The Effects of PET-Lined and Epoxy-Coated Lead and Copper Service Lines on Metals Leaching, Total Organic Carbon, and Chlorine Residual in Drinking Water. M.S. thesis, Master of Science in Environmental Engineering, Dept. of Civil Environmental and Architectural Engineering, University of Kansas, Lawrence, Kansas. Available from ProQuest Dissertations and Theses Global (Order No. 1571835; www.proquest.com/).

Bremer, P.J., B.J. Webster, and D.B. Wells. 2001. Biocorrosion of copper in potable water. *Journal AWWA*, 93(8):82–91.

Brown, R., and D. Cornwell. 2015. High-velocity household and service line flushing following LSL replacement. *Journal AWWA*, 107(3):E-140.

Brown, R., N. McTigue, and D. Cornwell. 2015. *Controlling Lead in Drinking Water*. Report 4409. Denver, Colo.: Water Research Foundation.

City of Naperville. 2002. *Design Manual for Public Improvements*. Naperville, Ill.: City of Naperville.

Clement, J., M. Hayes, P. Sarin, W.M. Kriven, J. Bebee, K. Jim, M. Becket, V.L. Snoeyink, G.J. Kirmeyer, and G. Pierson. 2002. *Development of Red Water Control Strategies*. Denver, Colo.: AwwaRF and AWWA.

Cook, D.M., P.S. Husband, and J.B. Boxall. 2015. Operational management of trunk main discolouration risk. *Urban Water Journal*, doi: 10.1080/1573062X.2014.993994

Cornwell, D.A., R. Brown, and S. Via. 2016. National survey of lead service line occurrence. *Journal AWWA*, 108(4).

DC Water. 2013. *Get Ready—We're Replacing the Pipes on Public Property*. Washington, D.C.: DC Water.

Deb, A.K., J.K. Snyder, J.O. Hammell Jr., E. Tyler, L. Gray, and I. Warren. 2006. *Service Life Analysis of Water Main Epoxy Lining*. Project 2941. Denver, Colo.: AwwaRF.

DeCarlo, E. 2004. Dielectric isolation concepts in water industry pipeline corrosion control. In *ASCE Pipelines*. Washington, D.C.: American Society of Civil Engineers. pp. 78, 146.

Del Toral, M., A. Porter, and M. Schock. 2013. Detection and evaluation of elevated lead release from service lines: A field study. *Environmental Science and Technology*, 47, 9300–9307.

Douterelo, I., R.L. Sharpe, and J.B. Boxall. 2013. Influence of hydraulic regimes on bacterial community and structure in an experimental drinking water distribution system. *Water Research*, 47:503–516.

Ellison, D., S. Duranceau, S. Ancel, G. Deagle, and R. McCoy. 2003. *Investigation of Pipe Cleaning Methods*. Denver, Colo.: AwwaRF and AWWA.

Friedman, M., and A. Blain. 2014. Use of monitoring, bench-scale, and pip rig studies to solve a pipe destabilization mystery. In *Proceedings of the AWWA Water Quality Technology Conference*. Denver, Colo.: AWWA.

Friedman, M., A. Hill, K. Martel, D. Holt, S. Smith, T. Ta, C. Sherwin, D. Hiltebrand, P. Pommerenk, Z. Hinedi, and A. Camper. 2003. *Establishing Site-Specific Flushing Velocities*. Denver, Colo.: AwwaRF and AWWA.

Friedman, M., A. Hill, S. Reiber, R. Valetine, G. Larsen, A. Young, G. Korshin, and C. Peng. 2010. *Assessment of Inorganics Accumulation in Drinking Water System Scales and Sediments*. Denver, Colo.: Water Research Foundation.

Friedman, M., B. Lakin, and R. Moore. 2012. Best practices cleaning mains: Clean, pig, or dig. *Opflow*, 38(8):10–13.

Friedman, M., K. Ervin, and R. Moore. 2014. The new Ice Age: Pigging effectively cleans water and wastewater pipelines. *Opflow*, 40(4):14.

Friedman, M., A. Hill, S. Booth, M. Hallett, L. McNeill, J. McLean, D. Stevens, D. Sorensen, T. Hammer, W. Kent, M. DeHaan, K. MacArthur, and K. Mitchell. 2015. *Metals Accumulation and Release Within the Distribution System: Evaluation of Mechanisms and Mitigation*. Denver, Colo.: Water Research Foundation.

Friedman, M., A. Hill, S. Booth, M. Hallett, L. McNeill, J. McLean, D. Stevens, D. Dorensen, T. Hammer, W. Kent, M. De Haan, K. MacArthur, and K. Mitchell. 2016. *Metals Accumulation and Release Within the Distribution System: Evaluation and Mitigation*. Denver Colo.: Water Research Foundation.

Guo, Q., P.J. Toomuluri, and J.O. Eckert Jr. 1998. Leachability of regulated metals from cement–mortar linings. *Journal AWWA*, 90(3):62.

Health Canada. 2009. *Guidance on Controlling Corrosion in Drinking Water Distribution Systems*. Ottawa, Ont.: Health Canada, Water, Air and Climate Change Bureau, Healthy Environments and Consumer Safety Branch.

Hill, A. 2001. Flushing guidelines. In *Proceedings of the AWWA Distribution System Symposium*. Denver, Colo.: AWWA.

Hill, A., M. Friedman, S. Reiber, G. Korshin, and R. Valentine. 2010. Behavior of trace inorganic contaminants in drinking water distribution systems. *Journal AWWA*, 102(7):107.

Hill, A., M. Johnson, T. Brewer, and R. Moore. 2014. Recognizing the distribution system asset/water quality management nexus. *Opflow*, 40(12):10.

Hock, V., E. Kirstein, K. Smothers, and J. Overmann. 1999. *Laboratory Evaluation of an In-Situ Coating Process for Mitigation of Lead and Copper in Drinking Water*. USACERL Technical Report 99/39. April. Washington, D.C.: US Army Corps of Engineers.

Hossain, A., J. Naser, and M.A. Imteaz. 2011. CFD investigation of particle deposition in a horizontal looped turbulent pipe flow. *Environmental Modelling and Assessment*, 16, 359–367.

Husband, P.S., and J.B. Boxall. 2010. Field studies of discolouration in water distribution systems: model verification and practical implications. *Journal of Environmental Engineering*, 136, 86–94.

Kirmeyer, G.J., G. Boyd, N. Tarbet, and R. Serpente. 2000. *Lead Pipe Rehabilitation and Replacement Techniques*. Report 90789. Denver, Colo.: AwwaRF.

LeChevallier, M.W., M.C. Besner, M. Friedman, and V.L. Speight. 2011. Microbiological quality control in distribution systems. In *Water Quality and Treatment: A Handbook on Drinking Water*, 6th ed. Edited by J.K. Edzwald. Denver, Colo.: AWWA.

Lytle, D., T. Sorg, and C. Frietch. 2004. Accumulation of arsenic in drinking water distribution systems. *Environmental Science and Technology*, 38(20):5365–5372.

McFadden, M., R. Giani, P. Kwan, and S. Reiber. 2011. Contributions to drinking water lead from galvanized iron corrosion scales. *Journal AWWA*, 103(4):76–89.

NFPA (National Fire Protection Association). 2011. *National Electric Code*, NFPA 70. Quincy, Mass.: NFPA.

NRC (National Research Council of the National Academies, Water Science and Technology Board). 2006. *Drinking Water Distribution Systems: Assessing and Reducing Risks*. Washington, D.C.: National Academies Press.

NRCC (National Research Council of Canada). 2013. Revisions and errata. In *2010 National Plumbing Code of Canada (NPC): 2013 Revisions Package*. Ottawa, Ont.: NRCC. Available at: www.nrc-cnrc.gc.ca/obj/doc/2010_npc_revisions_november2013.pdf

NSF International. 2009. NSF/ANSI Standard 53. *Drinking Water Treatment Units—Health Effects*. Ann Arbor, Mich.: NSF International.

NSF International. 2015. NSF/ANSI Standard 61. *Drinking Water System Components—Health Effects*. Ann Arbor, Mich.: NSF International.

Providence Water, 2013. Partial Lead Service Replacement Advisory. Providence, R.I.: Providence Water. Available at: www.provwater.com/projects/present/lsr/part_repl.pdf (accessed Aug. 27, 2013).

Randtke, S.J., C. Adams, E. Peltier, R. Lane, R. Carter Jr., Z. Breault, and J. Roberson. 2014. Evaluation of water service line lining and coating technologies. In *Proceedings of the Annual Conference of the American Water Works Association*. Denver, Colo.: AWWA.

Reiber, S. 2004. Disinfection Byproducts vs. Corrosion: A Case Study on the DC Water Lead Experience. Presentation to USEPA, April 2004.

Renner, R. 2006. Experiment confirms chloramine's effect on lead in drinking water. *Environmental Science and Technology*, 40(10):3129–3130.

Schock, M. 1989. Understanding corrosion control strategies for lead. *Journal AWWA*, 81(7):88.

Schock, M. 1999. Internal corrosion and deposition control. In *Water Quality and Treatment: A Handbook of Community Water Supplies*, 5th ed. Denver, Colo.: AWWA.

Schock, M., and R. Giani. 2005. Oxidant/Disinfectant chemistry and impacts on lead corrosion. In *Proceedings of the AWWA Water Quality Technology Conference*. Denver, Colo.: AWWA.

Schock, M.R., I. Wagner, and R.J. Oliphant. 1996. Corrosion and solubility of lead in drinking water. In *Internal Corrosion of Water Distribution Systems*, 2nd ed. Denver, Colo.: AwwaRF and AWWA.

Schock, M.R., R. Hyland, and M.M. Welch. 2008. Occurrence of contaminant accumulation in lead pipe scales from domestic drinking-water distribution systems. *Environmental Science and Technology*, 42(12):4285–4291.

Schock, M., A. Cantor, S. Triantafyllidou, M. Desantis, and K. Scheckel. 2014. Importance of pipe deposits to Lead and Copper Rule compliance. *Journal AWWA*, 106(7):E-336.

Speight, V., I. Douterelo, R. Sharpe, K. Fish, and J. Boxall. 2013. Impact of hydraulic events on organic and inorganic material in distribution systems. In *Proceedings of the AWWA Water Quality Technology Conference*. Denver, Colo.: AWWA.

Trussell, R.R., and I. Wagner. 1996. Corrosion of galvanized pipe. In *Internal Corrosion of Water Distribution Systems*, 2nd ed. Denver, Colo.: AwwaRF and AWWA.

USEPA (US Environmental Protection Agency). 2002. *Deteriorating Buried Infrastructure Management Challenges and Strategies*. Washington, D.C.: USEPA, Office of Water.

USEPA. 2006. Inorganic Contaminant Accumulation in Potable Water Distribution Systems (Total Coliform Rule Issue Paper). Washington, D.C.: USEPA.

USEPA. 2011. *SAB Evaluation of the Effectiveness of Partial Lead Service Line Replacements.* EPA-SAB-11-015. Washington, D.C.: USEPA.

Van Thienen, P., J.H.G. Vreeburg, and E.J.M. Blokker. 2011. Radial transport processes as a precursor to particle deposition in drinking water distribution systems. *Water Research,* 45, 1807–1817.

Wang, Y., H. Jing, V. Mehta, G.J. Welter, and D.E. Giammar. 2012. Impact of galvanic corrosion on lead release from aged lead service lines. *Water Research,* 46:5049–5060.

Wang, Y., V. Mehta, G.J. Welter, and D.E. Giammar. 2013. Effect of connection methods on lead release from galvanic corrosion. *Journal AWWA,* 105(7):E337–E351.

Welter, G., D. Giammar, Y. Wang, and A.F. Cantor. 2013. *Galvanic Corrosion Following Partial Lead Service Line Replacement.* Denver, Colo.: Water Research Foundation.

Chapter 6

Tools to Monitor Effectiveness of Corrosion Control and Metals Release Remediation

Abigail F. Cantor, Process Research Solutions, LLC
Christopher P. Hill, Arcadis
Richard Giani, CH2M

INTRODUCTION

Chapter 2 of this manual described the various mechanisms that can cause the release of metals into drinking water. As discussed in chapter 3, the various mechanisms of metals release can be indicated by measurement of specific water quality parameters. The chapter then explained how to use those parameters for an investigation of a water system to determine which mechanisms of metals release have the most significant influences in a specific system.

Chapters 4 and 5 described how to implement control of metals release targeted toward the various mechanisms.

This chapter moves the discussion forward (Table 6-1) and asks: Now that action is being taken to control metals release, how can the effectiveness of the action be verified?

Table 6-1 Chapter 6 key points

- A flexible and comprehensive monitoring strategy to verify treatment effectiveness and identify negative side-effects can begin by grouping water quality factors to correspond to the three main categories of metal release factors described in chapter 2.
- Because all water quality scenarios occurring in a distribution system cannot be studied, sampling locations in the distribution system should define the extremes of distribution system water quality. The entry points to the distribution system, distribution system high-water-age locations, water source blending fronts, neighborhoods of critically aged plumbing materials, and locations of repeated customer complaints typically exhibit the extremes of water quality in the distribution system.
- Additional sampling should be performed at a subset of individual buildings in the distribution system that are known to have water quality issues to better understand localized factors that influence the issues, such as existing pipe wall scales and piping configuration.
- Additional sampling should also be performed at existing distribution system sampling sites. Field tests on flowing water at those sites can indicate geographical locations of degraded water quality.
- Distribution system monitoring devices that relate lead and copper release to water quality parameters are variations of the early AwwaRF (Awwa Research Foundation, now called Water Research Foundation) pipe loop apparatus.
- A technique to make quick sense of monitoring data is the Shewhart control chart.
- An alternative or supplemental monitoring program should also be considered if more suitable to the specific water system and situation. Every monitoring program must consider monitoring strategy with selection of water quality parameters, choices of critical and representative sampling sites, and a method to analyze and make use of the data collected.

Is the action negatively affecting other aspects of water quality? Can water quality be further improved?

MONITORING STRATEGY

To answer the previously posed questions, a monitoring strategy that can be carried out routinely should be developed. Many strategies can be conceived. In some cases, use of multiple strategies may be beneficial.

One monitoring strategy begins by grouping mechanisms of metal release into three main categories of factors that shape distribution system water quality (Cantor et al. 2012), as was discussed in chapter 2:

1. Factors related to by-products of uniform corrosion of metals
2. Factors related to the biostability and microbiological activity in the water
3. Factors related to metal release by accumulation and transport

Water quality parameters that can quantify and track trends in these categories of factors are shown in Table 6-2. Some water quality parameters are helpful in describing more than one category of water quality–shaping factors. (Water quality parameters and their analyses are described in chapter 3.)

Water quality parameters selected from Table 6-2 for a specific water system's monitoring plan should be justified based on the known source water characteristics and water treatment chemicals. Some water systems may require other parameters not listed here. For example, a list of possible water quality parameters can be prepared that will aid in describing the three main categories of factors that shape a specific distribution system's water quality.

More details of monitoring plan development are presented elsewhere (Cantor 2009, Cantor et al. 2012). Alternative monitoring approaches are presented later in this chapter.

Table 6-2 Water quality parameters related to categories of factors that shape water quality in distribution systems

By-products of Uniform Corrosion of Metals	Biostability	Other Inorganics That Affect Accumulation and Release of Metals
Total alkalinity	Free chlorine (if using for disinfection)	Turbidity
Total hardness	Monochloramine (if using for disinfection)	Conductivity
Total dissolved solids	Total chlorine	Total and dissolved iron
Conductivity	Assimilable organic carbon and/or total and dissolved organic carbon	Total and dissolved manganese
pH	Ammonia-, nitrate-, nitrite-nitrogen	Total and dissolved aluminum
Temperature	Total phosphorus	(Optional) Metals scan for total and dissolved metals—list will vary based on laboratory and water system characteristics
Dissolved oxygen	Oxidation–reduction potential	Total and dissolved lead
Chloride		Total and dissolved copper
Sulfate		
Oxidation–reduction potential		
Orthophosphate		
Silicate		

SAMPLING LOCATIONS

To continue developing the monitoring plan, critical sampling sites must be identified. Water distribution system water quality changes over geographical location and time; it is quite difficult to obtain water samples that will be representative of its true nature. All of the individual circumstances cannot be studied. Instead, a combination of three strategies can be used.

Capturing the Water Quality Extremes of a Distribution System

The water quality characteristics of a distribution system can be bracketed between two extremes—between the characteristics of the freshest water and the water that has resided in the water system the longest time, or equivalent severe conditions.

Entry points to the distribution system. On one extreme, the freshest water with the highest concentrations of treatment chemicals can be found at the entry points to the distribution system. The effectiveness of the treatment chemicals are assumed to be at a maximum at these points.

Some water systems have only one entry point to the water distribution system; this is common for a surface water system with one treatment facility. Other water systems, such as groundwater systems with multiple wells, can have many entry points.

High water age locations. The extremes of water quality parameters opposite to the entry point of a distribution system are locations of high water age in the distribution system. *Water age* is another term for *residence time of water.* High water age locations occur at the farthest reaches of a distribution system, at dead ends, and in areas of low water usage. As water age increases, both added and naturally occurring chemicals in the water have more time to interact chemically and microbiologically. Many water quality issues in distribution systems can occur at high water age locations—lead and copper release, disinfection by-product formation, compliance issues with pathogenic microorganisms such as with the Total Coliform Rule in the United States, and discolored water. A strategic monitoring plan should address all of these issues because inter-relationships exist between them.

Blending fronts. Other critical sampling locations in water distribution systems occur where different water qualities blend together. This would be the case in a system with multiple water sources, such as in systems with multiple wells or in systems where groundwater supplements a surface water source. At these locations, water quality varies with multiple sources. The water quality swings can cause destabilizing conditions to existing chemical equilibriums and pipe wall scales. These swings, in turn, can impact corrosion and metals release.

Locations related to age of plumbing materials. Areas of high release of metals in a distribution system may be related to the age of the buildings in a neighborhood and the types of plumbing materials prevalent at the time the neighborhood was constructed. An inventory of system piping materials and age may provide better insight into selection of monitoring locations.

Locations of customer complaints. Customer complaints of discolored water, bad taste, or bad odor can identify locations in the distribution system where corrosion of metals along with other water quality issues may be occurring. Such locations, especially if geographical patterns of complaints are evident, should also be considered when selecting critical sampling sites for routine monitoring.

Capturing Water Quality at Critical Buildings in the Distribution System

In addition to knowing the possible best and worst water quality characteristics in a distribution system, it is important to understand how actual building plumbing responds to the system's water characteristics. Buildings where critical water quality characteristics have already been identified can be selected for individual study. For example, residences with known high lead concentrations that have been identified in Lead and Copper Rule compliance sampling can be studied in more detail to identify factors possibly controlling the release of lead. Piping configuration, water usage, water quality parameters such as those listed in Table 6-2, and lead concentration as it changes from the kitchen tap to the water main can add insight into lead release for that building. This type of sampling can also identify where the building and other critical buildings might fit into the extreme water quality bracket identified at sites mentioned in the previous section, "Capturing the Water Quality Extremes of the Distribution System."

Capturing Water Quality Throughout the Distribution System

Most drinking water systems have sampling sites identified throughout the distribution system because of regulatory or operations requirements. For example, some water systems that use chloramine disinfection have distribution system sampling sites for routine measurement of nitrification parameters to identify this common problem of chloramine use. Many water systems have distribution system sampling sites for regulatory monitoring for the presence of pathogenic microorganisms such as for the Total Coliform Rule in the United States. Other water quality parameters can be taken at those sites during regular visits and collected to describe the characteristics of water quality over time throughout the distribution system. Tracking field parameters such as conductivity, turbidity, and disinfection concentration can give immediate indications of lower water quality at a geographical location: Increased conductivity indicates a higher concentration of dissolved solids in the water; increased turbidity indicates a greater quantity of particulate matter in the water; decreased disinfection concentration indicates uptake by excessive growth of microorganisms and corrosion debris.

TYPES OF SAMPLES

With sampling sites selected, it must be determined whether water quality parameters should be measured in flowing water or in stagnating water. To understand when each type of sample is used, water pipes can be thought of as "batch reactors." A batch reactor is a vessel where chemicals are added and a reaction period allowed. At the end of a reaction period, the reacted product is removed from the vessel. In water pipes, freshwater flows into the pipe. The characteristics of the flowing freshwater can be sampled and analyzed. Chemical and microbiological interactions occur during the time that the water is in the pipe. If water is allowed to stagnate for a specific time, it can be sampled and analyzed for its characteristics to determine the outcome of the reactions. Examples of reaction outcomes in water are concentrations of lead, copper, and other metals and population of microorganisms.

Flowing Water Samples

Samples of flowing water should be obtained from sampling locations to define and compare general water characteristics along with treatment chemical dosages. For example, a water system adding free chlorine for disinfection and orthophosphate for lead and copper control may benefit from routinely monitoring flowing water for pH, temperature, conductivity, turbidity, oxidation–reduction potential (ORP), free chlorine, total chlorine, and orthophosphate. All of these parameters can be obtained with either field test equipment or on-line sensors. These parameters are descriptors of one or more of the categories of water quality factors as summarized in Table 6-2. The frequency of sampling can be quite often because no special laboratory analyses are required. Daily or multiple times a day may be beneficial at first to determine how often and how much each parameter varies over time. If a parameter does not vary greatly over a period of time, then once-a-week sampling can be sufficient. On-line measurement of these parameters provides the best understanding of variance and the greatest flexibility of sampling frequency.

Laboratory testing of other parameters suggested in Table 6-2, such as alkalinity, may be beneficial. As with the field test parameters, the frequency of sampling should be based on the nature of a parameter's variation. It is necessary to sample more frequently at first until the variation of a parameter is understood and an adequate frequency can be set.

Budget and labor availability for monitoring may force compromises in data-gathering frequency and number of parameters.

Stagnating Water Samples

For lead and copper control, it is important to relate the flowing water parameters to lead and copper release as described above in the discussion regarding batch reactors. Residential plumbing has functioned as a batch reactor in many regulatory and investigative sampling schemes, where water is allowed to stagnate in the piping before samples are taken.

However, it is not always practical to routinely sample in buildings for lead and copper release, so an alternative model for building plumbing is needed. Such alternatives are discussed below.

AwwaRF pipe loops. A pipe loop apparatus was developed by the Water Research Foundation (formerly called the Awwa Research Foundation, or AwwaRF) in anticipation of the 1991 Lead and Copper Rule (EES 1990). The original apparatus consisted of one or

Courtesy of the US Army Corps of Engineer's Washington Aqueduct.

Figure 6-1 Flow-through AwwaRF pipe loop apparatus

more lengths of lead or copper pipe, where each pipe would be able to hold at least 1 L of water so that a 1-L sample, similar to a United States Lead and Copper Rule residential compliance sample volume, could be obtained (Figure 6-1). Pipes in the test apparatus should be harvested for lead pipes, and new or old pipes should be used for copper pipes. The importance of a harvested pipe is that it maintains the current condition of the pipe wall scales. The pipes are attached to nonreactive plastic pipe that connects to fresh distribution system water. Flowmeters control the flow to each pipe. Water flows from the distribution system through the pipes and to waste, as occurs in building plumbing. The water flow from the distribution system is controlled by a programmable timer controlling a valve to turn water on and off, as is the water usage pattern in a building.

The timer also allows for a period of stagnation, as described above regarding reaction time in batch reactors. Because the US Lead and Copper Rule calls for a minimum of six hours of stagnation in residential plumbing before sampling, six hours is a common stagnation time used before sampling a pipe loop apparatus.

The original AwwaRF pipe loop apparatus was only used for corrosion control treatment chemical comparisons of entry point water; it was not used for routine monitoring at various locations around a distribution system. But even if it would have been used for distribution system monitoring, placing it out in the water system would have been very awkward because of its size.

Also, when using new pipes, the role of pipe films and scales on existing system piping is not captured. When old pipes harvested from water distribution systems are used, issues remain; there is no control over how representative each harvested pipe may or may not be in the distribution system. In addition, metal particulates from the debris on the old pipe walls get disturbed and interfere with the metal concentration data for a long period of time (AwwaRF and DVGW-TZW 1996).

Mini pipe loops. Mini pipe loops were developed to utilize the AwwaRF pipe loop concept as a means to monitor around a water distribution system (Cantor et al. 2000). They were originally constructed in 1996 at one-quarter the size of a standard pipe loop

Courtesy of Process Research Solutions, LLC, Madison, Wis.

Figure 6-2 Mini pipe loop used as a distribution system monitoring station in 1996

Courtesy of Process Research Solutions, LLC, Madison, Wis.

Figure 6-3 Mini pipe loops of three metals for chemical treatment comparison in 1999

and were skid-mounted. While the full size apparatus might take up a large area at a water treatment plant, as the one shown attached to a wall in Figure 6-1, the mini pipe loops were portable to about a 3-ft by 5-ft (91-cm by 152-cm) floor space in the distribution system, as shown in Figure 6-2. The copper mini pipe loops were placed around a distribution system to monitor the water for effects of an added chemical that was thought to be increasing the copper levels in the water (Cantor et al. 2000; see Figure 6-2). This strategy successfully identified the chemical as the major influence of the copper problem and also identified when residences could be resampled to show lower copper concentrations after removal of the chemical. In 1999, mini pipe loops were constructed for a research project to test the effects of chlorine addition on corrosion (Cantor et al. 2003; see Figure 6-3). Three metals and three chemical treatments were studied using nine pipe loops. The complete apparatus fit into the corner of a small well house.

However, the apparatuses used new pipe material, and the interaction of existing pipe films and scales with the water still could not be captured. This is an important limitation that must be taken into consideration when evaluating data from these apparatuses.

Process Research Solutions monitoring stations. After the experiences with the mini pipe loops, it became obvious that a distribution system monitoring station used as a routine gauge of water quality adds much insight into lead and copper and other metals release into water during routine water system operations and through system operational changes. However, a distribution system monitoring station must meet the following criteria:

- Is relatively inexpensive
- Is easy to move from location to location
- Does not take up excessive space
- Is cushioned and isolated from vibrations that can interfere with pipe accumulation release
- Is straightforward to operate
- Can be used for either off-line chemical treatment comparison tests or as a gauge of water quality around a distribution system
- Allows for repeat collection of water samples
- Is easily accessible to water utility staff
- Has a uniform configuration
- Has uniform water flow rate and water usage
- Has a steady pressure
- Is capable of water stagnation for a uniform time
- Captures routine operation of the water system as well as treatment and operational changes that may occur over time
- Captures the interaction of pipe scales and films with the water
- Makes pipe film and scale analysis possible and economical so that the young scales developed during the monitoring period can be compared to older existing scales on actual water system pipe
- Does not destroy pipe film and scale during the metal sample preparation

To meet these criteria, the mini pipe loops were modified in 2006 into what is called a Process Research Solutions (PRS) monitoring station (Cantor 2009; see Figure 6-4). Instead of a piece of coiled pipe, stacks of metal plates that are secured in a section of larger plastic pipe are exposed to system water. The surface area of the metal plates to the volume of water held in the test chamber is similar to a 1.77-in.-diameter pipe used as a pipe loop. This equivalent diameter pipe was the most economically and physically practical to achieve. Figure 6-5 shows the metal plates stacked inside the test chambers before the chambers are sealed.

The plates do *not* function as metal coupons as described later in this chapter in the "Coupon Studies" section. In coupon studies, the metal plates are cleaned of corrosion debris and only coupon weight loss is considered as an indicator of corrosion. With the PRS monitoring station, the metal plates are located in a test chamber to provide an equivalent metal surface area exposed to water as would be experienced in a pipe; the plates create an abstraction of a pipe for a convenient testing configuration. Lead and copper concentrations released from the metal plates are obtained in water samples stagnating adjacent to the metal plates, similar to a water sample from a pipe in a building. At the end

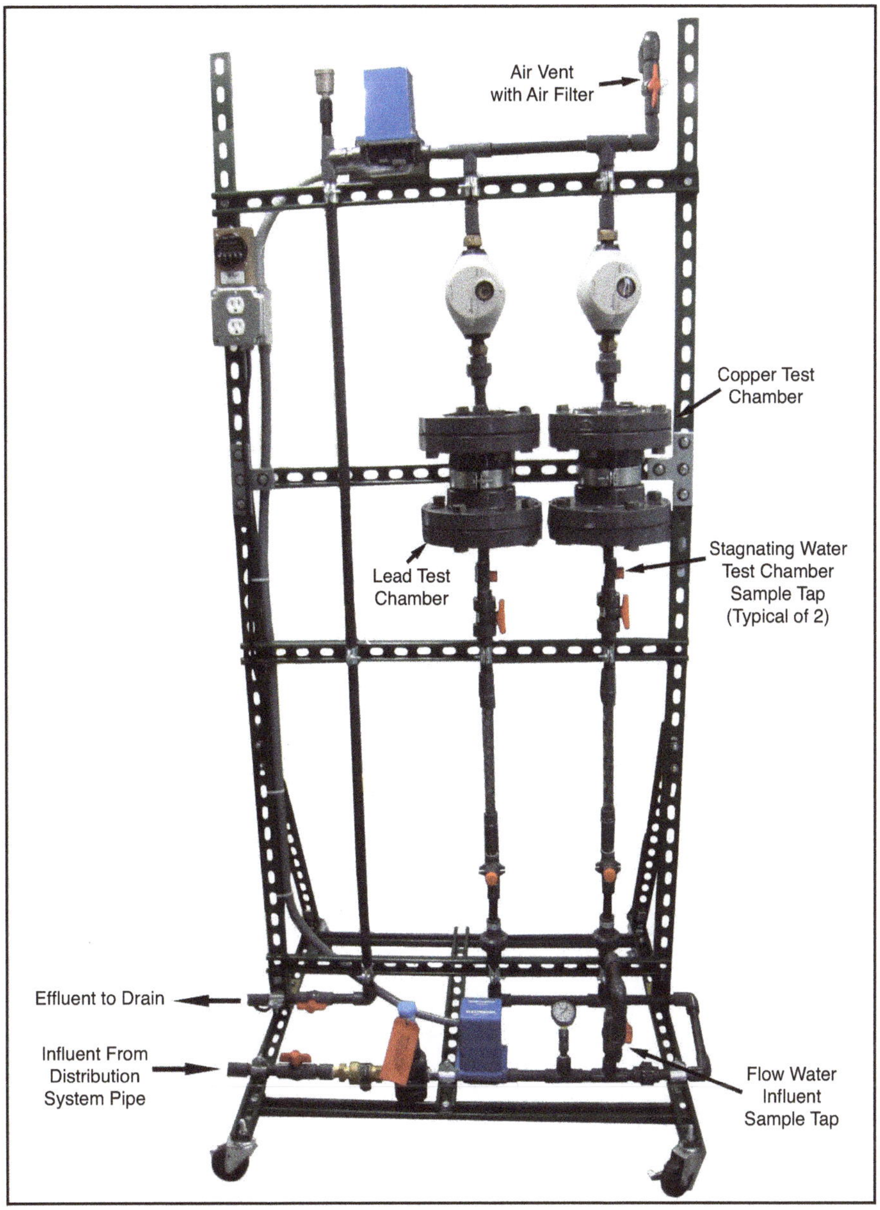

Courtesy of Process Research Solutions, LLC, Madison, Wis.

Figure 6-4 PRS monitoring station, a standardized distribution system monitoring station

of a monitoring period, the metal plates are extracted from the test chambers with their surface corrosion debris intact. The layers and composition of the corrosion debris are studied in the same way that a pipe can be harvested from a distribution system and its pipe wall debris examined. The compounds as layered on pipe walls or the PRS monitoring station metal plates give insight into the chemistry that is possible in the specific water system and into what shapes the water quality in that water system.

The flow characteristics through a pipe configuration versus through a test chamber filled with flat plates are different, and the flow characteristics can influence the type and quantity of debris that builds up on metal surfaces. However, this fact does not take away from the significance of the data from a PRS monitoring station. One reason is that the consistency of obtaining water samples from a test chamber is a major benefit of using a

Courtesy of Process Research Solutions, LLC, Madison, Wis.

Figure 6-5 View of plates sitting inside and stacked in open test chambers in a PRS monitoring station

standardized test apparatus versus obtaining water samples from buildings or pipe loops with varying configurations. Another reason is that the PRS monitoring station operational parameters are designed to exaggerate the release of lead and copper with more stagnation than flow where the lead and copper release data typically represent a maximum release of lead and copper possible in a specific water system.

The PRS monitoring station is operated in the same way that standard or mini pipe loops are operated. A timer controls the flow of water through the device as well as the time of stagnation. First-draw water samples are taken from the test chambers at the end of a prescribed stagnation period, collecting water that has been exposed to the metal surfaces. Flowing water samples of water that fill the test chambers can be taken from the influent sample tap.

Water quality parameters analyzed in the test chamber stagnation water are typically the total and dissolved metal corresponding to the type of metal in the test chamber. That is, for a test chamber with lead plates, lead analyses are performed; for copper plates, copper analyses are performed; for brass plates, both lead and copper analyses are performed. Alternatively, a scan of 15 to 25 metals can be performed by some laboratories. That information can be quite interesting for each test chamber, no matter what type of plates are installed.

Another parameter of interest for the stagnating water from a PRS monitoring station is the population of microorganisms that become entrained in the water from the biofilm that has developed on the metal plates' surfaces during the stagnation time. The population of microorganisms is best measured by the adenosine triphosphate (ATP) test. DNA tests can also be performed to determine what types of microorganisms have the potential to grow in the water system (refer to chapter 3). PRS monitoring station monitoring data show that different quantities and types of microorganisms grow on different types of metals (Cantor 2017).

Another advantage of the PRS monitoring station is that after a certain period of exposure time, the metal plates can be removed from the test chambers and analyzed for mineralogical, chemical, and microbial composition of the pipe film and scale. This is similar to the analyses performed on the interior walls of pipes that have been harvested from water distribution systems (refer to chapter 3). The thin, square PRS monitoring station metal plates need little preparation to be set into the analytical instruments for chemical analyses, and biofilms are easily extracted for analysis by ATP and DNA as well. The ability to easily analyze the inorganics and microbiological accumulations on the metal plates aids in relating the PRS monitoring station water quality data to older lead pipes in distribution systems and determining the copper release in younger copper pipes in the distribution system. The scales on the metal plates have been found to be "young," as would be expected for a pipe with the time of exposure allowed in the PRS monitoring station, typically six months to one year. The only disadvantage is the PRS monitoring station will not be able to mimic metals release from pipe scales that have developed over decades of time; therefore, some individual building sampling should be performed for comparison. Nevertheless, the plate analyses can give insight into oxide and carbonate formation, interference of other chemicals in scale development, chemical accumulation patterns on various types of metals, and biofilm formation on various types of metals.

Water Research Foundation Project 4317 pipe rig. The Water Research Foundation evaluated another pipe loop apparatus in 2011 in their Project 4317 with the intent of using it as a monitoring device around the distribution system (Parks et al. 2014). The desire was to use such a device as an alternative for residential sampling, similar to the function of the mini pipe loops and the PRS monitoring stations.

The Project 4317 pipe loop apparatus used three types of piping configurations:

1. 5 ft (152 cm) of ¾-in.-diameter copper tube with 12 in. (30.5 cm) of 50:50 lead:tin solder melted near the effluent end of the tube
2. 2.5 ft (76 cm) of ¾-in.-diameter polyvinyl chloride (PVC) pipe with 2.5 ft (76 cm) of ½-in.-diameter CDA360 brass rod glued inside plus 2.5 feet (76 cm) of empty ¾-in.-diameter PVC pipe
3. 2.5 ft (76 cm) of ¾-in. copper tube galvanically connected to 2.5 ft (76 cm) of ¾-in. lead pipe

Nine pipes were in the apparatus with each of the three configurations in triplicate (Figure 6-6).

The Project 4317 apparatus was installed in five water utilities in the United States. Pipe loops in the distribution system can be used as a routine gauge of lead and copper transfer into water and "may provide a more consistent measure of a water's corrosivity and a utility's corrosion control effectiveness" (Parks et al. 2014). "Two participating utilities, which shared the same source water but had different water processes had different results for lead and copper leaching, reinforcing the significance of treatment differences and water chemistry in corrosion control" (Parks et al. 2014).

Summary criteria for pipe loop–style apparatuses. In summary, mini pipe loops (Cantor et al. 2000), PRS monitoring stations (Cantor 2009, 2017; Cantor et al. 2012), Project 4317 apparatuses (Parks et al. 2014), and similar devices are based on the original AwwaRF pipe loop apparatus (EES 1990) design and can serve as alternatives for residential plumbing monitoring as well as be used for off-line treatment chemical comparison tests. These are all devices connected directly to a water distribution system with water flowing through pipes to waste as it would in a building.

If chemical comparison tests are run using the apparatuses, they are typically run with the device at an entry point where water treatment is located. When chemicals are added for tests, they are dosed into the pipeline with a chemical dosing pump and injection port, just as would be done in an operating water system.

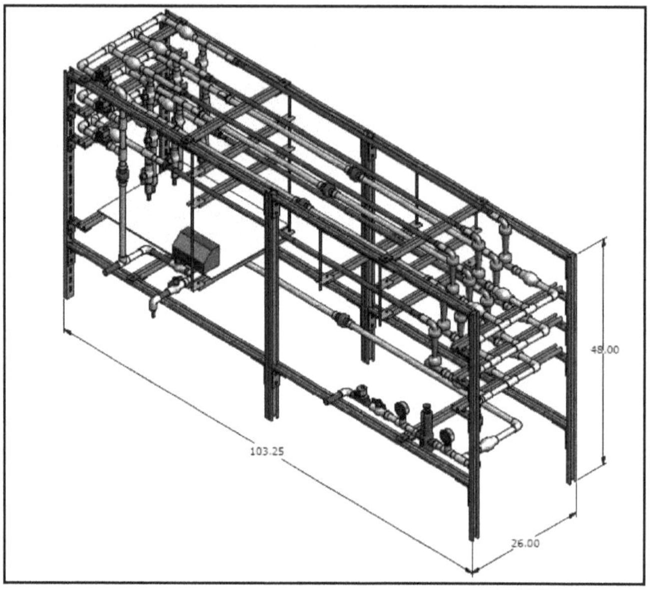

Courtesy of Water Research Foundation

Figure 6-6 Water Research Foundation Project 4317 pipe rig

Most important is that the monitoring device should be accompanied by a comprehensive and flexible monitoring strategy that can identify information representative of the specific water system. Critical sampling locations, water quality parameters, and sampling frequencies must be identified. The comprehensive monitoring effort must lead to uncovering mechanisms of metal release in a water system and relating this to other distribution system water quality parameters and issues (Cantor 2009).

COUPON STUDIES

Coupon testing is a type of analysis that came to the water industry from industrial applications. In industry, the corrosivity of process water or any process flow is determined by placing small metal samples (Figure 6-7) inside a pipe exposed to the fluid of interest for a set period of time. The weight of the coupon is measured before and after its exposure, and the weight loss is calculated as corrosion over time. This technique is sufficient for industries because their concern is determining when metal tank and pipe walls might lose strength or form holes because of metal loss. This does not necessarily translate to protection of consumers from minute quantities of metal in water. Nor does it necessarily account for other mechanisms of metal transport outside of uniform corrosion.

Important coupon testing criteria include the following:

- The coupon metal must be representative of the piping material of interest.
- The water chemistry of the coupon exposure must be indicative of the distribution system.
- The hydraulic regime across the coupon must reflect the flow regime across the distribution system pipe wall or plumbing appurtenance.
- The duration of the test must allow for development of corrosion scales and/or passivating films that influence the corrosion rate of the underlying metal.

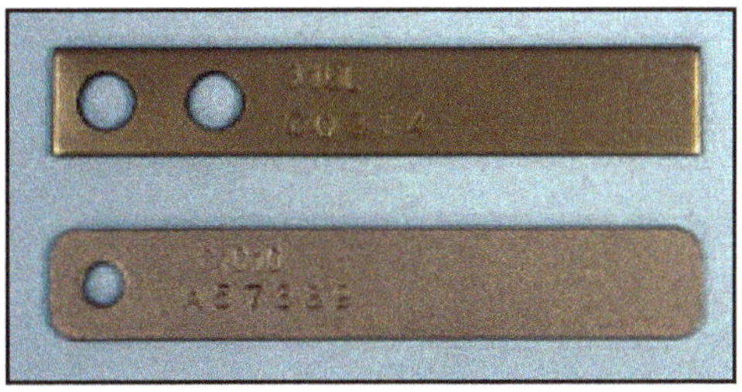

Courtesy of Metal Samples Company

Figure 6-7 Typical copper coupons

Figure 6-8 Typical coupon study apparatus

Coupon data are commonly presented as a penetration rate expressed in terms of unit area of mass loss per day of exposure (g-metal/cm^2/day). It is common, although not necessarily correct, to assume that metal loss is evenly distributed over the entire coupon test surface, making it possible to present the data as a uniform surface penetration rate. On metals that generally corrode uniformly (i.e., copper, copper alloys, Pb/Sn solders, etc.), it is acceptable to assume uniformity of metal loss and that uniformity be used to calculate an approximation of service life. However, this type of analysis may be inappropriate for pitting surfaces or other localized forms of corrosion where a uniform representation of metal loss is likely to grossly overestimate remaining service life. In such cases, penetration of the pipe wall will occur in a small fraction of the time predicted by a uniform assumption of corrosion.

As a direct measure of corrosion rate in a drinking water environment, coupon techniques are sometimes limited to those metals, such as iron and steel alloys, where the nominal corrosion rate produces sufficient mass loss to be readily measured. Under some exposure conditions, metals such as copper, copper alloys, and lead have such a low nominal corrosion rate that the mass loss cannot be precisely measured unless extended exposure periods are employed.

There is no single standard regarding coupon geometry, materials, or exposure protocols in drinking water systems. Although some coupon techniques have been developed specifically for drinking water distribution systems, others have been borrowed from different industries. ASTM International has certified several methodological variants for use in the evaluation of metal loss on coupon exposures. The most widely used technique (ASTM D2688-83 Method B, 1983a) relies on flat, rectangular coupon specimens mounted on nonmetallic stems and inserted directly into the flowstream of the pipe, usually at an elbow or tee (Figure 6-8). This technique can be used to make relative assessments of corrosion at different locations in a distribution system and for comparative analyses of corrosion inhibitors. However, because the hydraulic flow lines around a flat coupon positioned mid-pipe are substantially different from the flow lines at the pipe wall, this coupon technique may be inappropriate when a precise estimate of piping corrosion rates is required. Also, it is frequently difficult to obtain flat coupons that are truly representative of pipe materials.

Table 6-3 presents a comparative summary of published coupon protocols, including geometry and exposure conditions.

A rigorous coupon evaluation involves measuring weight loss over an extended period of time and requires coupon sacrifices at multiple points in the exposure cycle. A weight loss against time curve is drawn, and the corrosion rate at any point is the gradient of that curve at that exposure duration. Experience has shown that on most metal surfaces, corrosion rates change over the course of the exposure, with the highest corrosion rates occurring at the beginning of the exposure and then rapidly decreasing to a lesser and more constant rate. Hence, any comparison of corrosion rates for a particular metal must be standardized to specific exposure duration.

ALTERNATIVE MONITORING APPROACH

Another strategy for routine corrosion monitoring involves measuring key corrosion parameters (considered as "Level 1" parameters, similar to Level 1 parameters described in Table 3-3) at the same time and location that regulatory coliform samples are collected. These parameters would also be measured routinely at the entry point to the distribution system.

This technique is efficient in that it does not require extra time to get to the sample site because the sampler is already there to collect total coliform samples. Setting Level 1 parameters and goals is system specific and determined by each utility based on its treatment, corrosion control method, and other factors. For instance, Table 6-4 shows an example of Level 1 water quality parameters and respective target ranges as considered by a large surface water utility that adds alum, orthophosphate, and chloramine, and has hard water and a high percentage of cast-iron mains. Because of the high percentage of cast-iron mains, routinely monitoring color or turbidity is important to determine if corrosion issues are occurring from the mains. One must keep in mind that the water quality parameters are only indicators of degrading water quality and cannot determine if corrosion control of lead and copper has been achieved.

Level 1 parameters outside of their set target ranges in the distribution system would trigger follow-up monitoring the next time the coliform site is scheduled for coliform sampling.

Potential follow-up monitoring water quality parameters (Level 2 parameters) and their respective target ranges are listed in Table 6-5 for the same utility. Level 2 parameters are more decisive in indicating the potential reason why the key (Level 1) water quality parameters are outside of their range and if localized corrosion may be occurring.

If two parameters are outside their range during follow-up monitoring, follow-up action would be required. For example, low pH, elevated nitrite, and ATP may indicate microbiologically influenced corrosion while elevated iron and color may indicate metal accumulation in the main.

Depending on the issue, follow-up actions could involve localized unidirectional flushing or valve investigations (to identify if a valve was inadvertently closed). If Level 1 or Level 2 parameters continue to be outside target ranges, further actions such as main replacements/main rehabilitation, installing chlorine booster stations, minimizing water

Table 6-3 Summary of coupon protocols used for distribution system corrosion measures*

Protocol Name	Coupon Geometry	Mounting Arrangement	Exposure Duration	Comments	References
ASTM flat coupon standard Corrosivity of water in the absence of heat in transfer	Flat rectangular coupons stamped from sheet stock (0.5 in. × 4 in. [1.27 cm × 10.2 cm]). Both sides exposed to water flow.	Held on insulating stem and inserted into flowstream of distribution main at elbow or tee.	12 to 24 months.	Generally used to monitor corrosion processes on distribution system materials. Used primarily with mild steel or cast-iron materials. Useful for pitting evaluations.	ASTM D2688-83 Method B (1983a), ASTM G1-81 (1981), ASTM G46-76 (1976)
Illinois State Water Survey (ISWS) protocol machined nipple test	Machined pipe nipples (0.75 in. [1.9 cm] ID, 4 in. [10.2 cm] length). Exterior coated with epoxy resin (9.2 in.^{2} [60 cm^2] exposed surface area).	Fitted into a machined PVC pipe section and plumbed into bypass off distribution main.	Minimum 120 days recommended.	Coupon fabrication requires extensive machining. Coupon processing is chemically based (acid baths).	ASTM D2688-83 Method C (1983b)
Modified ISWS coupon sleeve tester (University of Washington)	Short pipe section (0.75 in. [1.9 cm] ID, 1.5 in. [3.8 cm] length). Exterior phenolic resin coated (3 in.^{2} [20 cm^2] exposed surface area).	Coupons gasketed and held in clear acrylic sleeve (up to 10 coupons per sleeve). Flow restricted to coupon interior. Easy coupon substitution. Flow controlled to duplicate alternating turbulent and stagnation conditions.	Coupons sacrificed at 30-, 90-, and 180-day intervals.	Coupon processing is mechanically based. Measurement precision improved by use of multiple coupon exposures. Can be used to monitor corrosion process on residential plumbing materials.	Reiber et al. (1988)
Water Research Centre, Swindon, United Kingdom, coupon rig	Flat rectangular coupons (3 in. × 0.5 in. × 0.15 in. [75 mm × 12.5 mm × 4 mm]; 3.8 in.^{2} [25 cm^2] exposed surface).	Coupons mounted on insulating stems and held in flow control rig.	Coupons sacrificed at periodic intervals for up to 22 weeks.	Developed for use with cast-iron coupons but suitable for use with other materials. Flow rig has high- and low-velocity sections.	Williams et al. (1984)
US Army Corps of Engineers Research Laboratory tester	Flat rectangular coupons similar to the ASTM D2688-83 standard.	Mounted on insulating stems in special pipe rig, designed to hold a large number of coupons under controlled hydraulic conditions.	Not specified.	Coupon surface exposure not typical of pipe wall hydraulics.	Temkar (1988), Temkar et al. (1987)

*ID = inner diameter; PVC = polyvinyl chloride

Table 6-4 Example of Level 1 water quality parameters

Parameter	Target
pH	≥7.2 and ≤7.8
Orthophosphate (total)	2.0 ± 0.5 mg/L as PO_4
Total chlorine	≥0.5 mg/L
Color (apparent) or turbidity	≤25 color units or ≤2 NTU

NTU = nephelometric turbidity units

Table 6-5 Example of Level 2 water quality parameters

Parameter	Target
pH	≥7.2 and ≤7.8
Orthophosphate (total)	2.0 ± 0.5 mg/L as PO_4
Orthophosphate (dissolved)*	2.0 ± 0.5 mg/L as PO_4
Color (apparent)	≤25 color units†
Iron (total)	≤0.3 mg/L‡
Ca hardness (total)	Not applicable
Ca hardness (dissolved)*	Not applicable
Calcium precipitate§	≤10 mg/L as $CaCO_3$
Aluminum (total)	≤0.1 mg/L
Nitrite	≤0.05 mg/L as N-NO_2
Adenosine triphosphate, or ATP	≤500 cATP ME/mL

*Filter using a 0.45-micron filter

†The target level for apparent color is *less than or equal to* 40 color units if sampling from a hydrant.

‡The target level for iron is *less than or equal to* 0.5 mg/L if sampling from a hydrant.

§Total Calcium Precipitation = Total Calcium Hardness – Dissolved Calcium Hardness

age, ensuring proper water velocities, and so on, should be considered. Figure 6-9 is an example of an action flow chart taken from the same utility.

Total coliform locations are efficient because they are placed throughout the distribution system, but they are not everywhere and in many cases they are not along dead-end lines. It is also recommended to include sites in places known to have a dead end or long retention times. In addition, some utilities will conduct the same type of monitoring during customer-complaint investigations where sampling is conducted in the home and at a nearby sampling station. If there is no sampling station, then fire hydrants can be used; however, proper flushing is required before taking the sample and wider target ranges are recommended as a result of metals accumulation in the hydrant service line. Table 6-5 also provides an example of target ranges for samples collected from hydrants. Target levels should be established based on the utilities' specific water quality and explicit needs.

DATA MANAGEMENT AND ANALYSIS

Data Management

Increased monitoring means increased quantities of data. Data must be stored properly to be kept convenient for use. The best way to store data is in a relational database. A

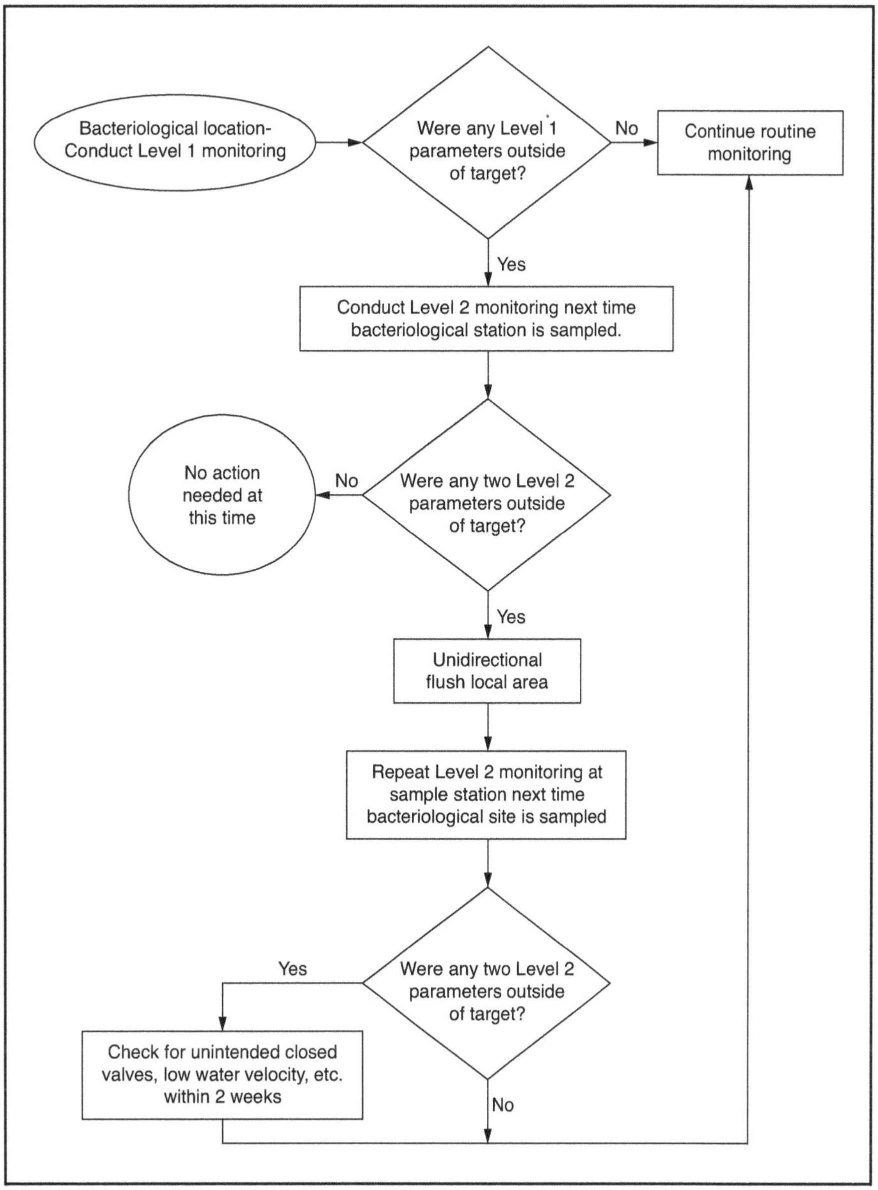

Figure 6-9 Process flow chart for multi-level corrosion control sampling at total coliform sampling sites

relational database allows one to retrieve data quickly using flexible selection criteria. Typically, the database file must be structured properly by a trained professional, and a user interface is needed to make data entry and retrieval user-friendly. Various software programs supply this database file structure and user interface. It is a good investment to purchase such a program that works in association with common database software.

Computer spreadsheet programs are for analyzing and reporting data. The most robust data management systems allow the user to input the data to a database file, retrieve selections of that data quickly, and transfer the data to a spreadsheet program for analysis. It is advisable for technically based people to have basic spreadsheet skills, such as setting up tables of data, copying and pasting rows and columns, and creating time series graphs. Some of these steps can be performed automatically using the features of many data systems commonly used at water utilities, such as Laboratory Information Management System (LIMS) and supervisory control and data acquisition (SCADA).

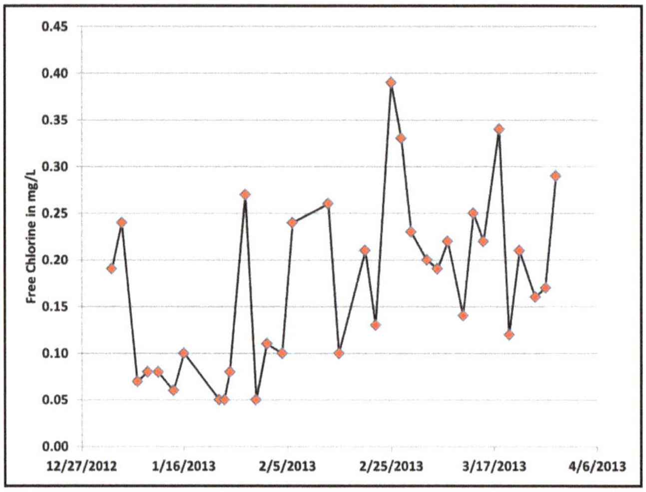

Figure 6-10 Example of a time series plot

Original data should not be stored in a spreadsheet program. However, if using database software is not practical for a water utility, then use of a spreadsheet program to store the data is acceptable. To keep the data organized and readily usable, a separate spreadsheet should be designated for each sampling site and each parameter. The date and time of sampling should be listed in one or two columns associated with a column of results for that parameter. The units of measurement should also be listed. Data analysis using the original data should be performed in a separate spreadsheet file (workbook) so that the original data file is used for data collection only.

All data files should be backed up properly so that a copy of the original data exists.

Introduction to Data Analysis Using Shewhart Control Charts

All water system data are taken over time, whether it is water quality data from a distribution system monitoring program or operational data, such as pounds of treatment chemical used or water flow rate. For data analysis, at the very least, these data should be plotted on time series graphs. Time series graphs use date or time on the x-axis and the unit of measurement for some parameter of interest on the y-axis. Figure 6-10 is an example of a time series plot where high and low values can be easily identified. These graphs can aid in seeing extreme values of a measurement and triggering an investigation into why they occurred, whether it was a sampling or measurement issue or an actual water system operations issue.

There is a method that can assist in determining which data points on a time series plot are too high and which are too low. It can also identify other water quality trends, indicating a changing water system. The method is to construct a Shewhart control chart. It is a data analytical technique that has been borrowed for water quality and water system data (Cantor et al. 2012) from the field of industrial process control and improvement called statistical process control. This technique was developed in the 1920s by Walter Shewhart of Bell Laboratories and was later integrated into many industries by W. Edwards Deming, a champion of process control and improvement (Wheeler and Chambers 1992).

At the center of the technique is the Shewhart control chart. The control chart is essentially a graph of data over time. Ease of interpreting the data is achieved by the guidelines plotted on the graph, which are plots of the Shewhart statistics calculated from the dataset. The statistics describe the expected variation of the data if the system remains operating under its existing conditions. Data points that fall outside of the lines of expected variation

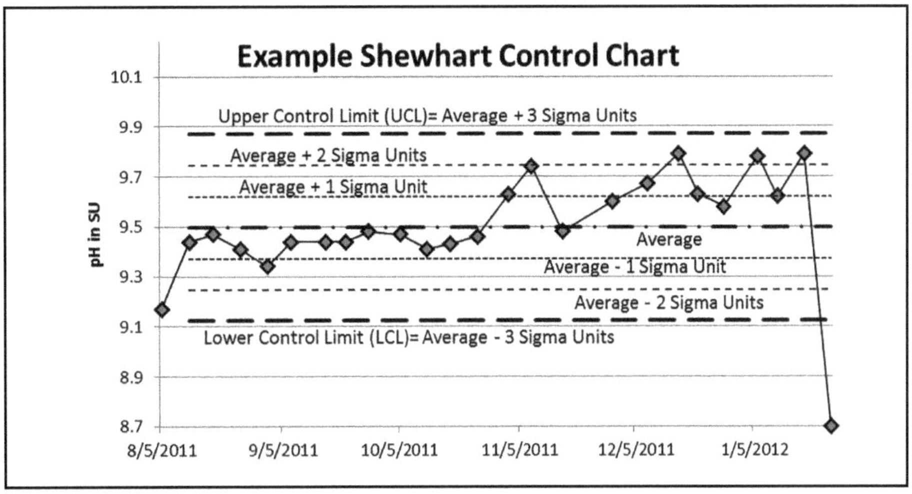

Figure 6-11 Summary of Shewhart control chart characteristics

show that new factors are influencing the system. Other data patterns on the chart can show that changes to the status quo are beginning to occur (Wheeler and Chambers 1992).

Figure 6-11 summarizes the control chart's features. For Shewhart control charts, monitoring data are plotted over time. The average is drawn as a line through the data. A unit of variation, called a sigma unit, is calculated and then used to define the range (±3 sigma units around the average) in which 99 percent of the data are expected to fall. Standard deviation is a type of sigma unit. However, standard deviation can only be used on data from randomized experiments where each data point is independent of each other. In water quality monitoring, samples can only be taken sequentially over time, like on an assembly line; the conditions at a previous sampling time might affect the conditions at the next sampling time. Therefore, the non-parametric sigma unit is used. In this discussion, the sigma unit is based on the difference between successive data points, called a moving range. The reader is referred to other references for a better understanding of the technique (Wheeler and Chambers 1992).

For Figure 6-11, the following are defined:

- Average + 3 sigma units = upper control limit = UCL
- Average – 3 sigma units = lower control limit = LCL
- Between the UCL and the LCL = the expected range of the data. This is the range that will result from routine factors working on the system.

Patterns of atypical data can be found on the graph when

- data fall outside the 3 sigma unit lines,
- at least two out of three successive values fall on the same side of the average and are 2 sigma units or greater away from the average,
- at least four out of five successive values fall on the same side of the average and are 1 sigma unit or greater away from the average, or
- eight or more successive points fall on the same side of the average line.

Any data following the above patterns should trigger an investigation of system operations to determine the cause. If results are unfavorable, corrections can be made to system operations. If results are favorable, the continuation of such operating conditions could be beneficial.

Using Shewhart's and Deming's statistical process control techniques do not require an in-depth background in statistics. The construction and use of a control chart is not complicated. Software packages that can create the charts are also available. However, it is important to understand certain nuances about the technique. First, the way that the Shewhart statistics are calculated will depend on the frequency of sampling for a specific parameter at a specific site. If products on an industrial assembly line were measured for certain characteristics every 15 minutes, the Shewhart control chart technique could be used to group the four parameters taken each hour and use their average value in the statistics calculation instead of their individual values. A water utility using an on-line sensor for measurement of a water quality parameter may also choose to group data taken every 15 minutes by hour or possibly group all measurements by day. The decision as to how big a data group should be averaged is up to the analyst. The decision is based on what questions are to be answered with the data and what degree of sensitivity is desired on the control chart. "Subgroups must match the time period which characterize the same changes in process" (Wheeler and Chambers 1992). For people who want to answer important questions for frequently sampled measurements, it is suggested that they read texts on the nuances of Shewhart control charts (Wheeler and Chambers 1992).

However, there are situations—especially in water systems—where frequent measurements cannot be made. This is true for the monitoring previously described. Data from a pipe loop–style apparatus in the distribution system are obtained infrequently—once a week, every other week, once a month. Data from some regulatory site sampling, such as for the Total Coliform Rule in the United States, are obtained infrequently at each regulatory site, from a few times a month to every few months. Therefore, much time passes between measurements along with changes that may be occurring in the distribution system that it does not make sense to group one data point with others; they all represent a different state of the distribution system. In this situation, data are not to be grouped. "Each value is uniquely identified with a specific period of time. This is a logical subgroup of size of n = 1" (Wheeler and Chambers 1992). Therefore, a chart of subgroup size n = 1 should be calculated.

Another nuance concerning Shewhart control charts is whether to use the *average* of data versus the *median*. Some argue that using the average of data creates a control chart with possibly very large variations of data that may not be representative of typical operations, and, therefore, median values should be used to smooth out the graphs (Cornwell et al. 2015). Others warn that "The chart for the subgroup medians will never be as sensitive as the regular control chart for subgroup averages. It will always be slower in responding to changes in the process. [The subgroup averages chart] should always be preferred over a median chart" (Wheeler and Chambers 1992). A high variation in a dataset is a signal to identify the influencing factors in a water system, whether it is a faulty sampling or analytical method or an actual system operations issue. Dismissing high variation as "noise" or "outliers" is a lost opportunity to understand and improve a process.

In the monitoring strategies described in this chapter, the goal is to respond to water quality trends and determine what is pushing the water system to a different state. It is true that sometimes the trends on a sensitive, average-based Shewhart control chart might be false alarms. However, by responding to each alarm, even seemingly minor issues, such as sampling protocols, problems with a chemical feed pump, or problems with operator protocols, can be corrected and those factors removed from keeping a parameter within statistical control in a distribution system.

Shewhart Control Chart Technique

Table 6-6 depicts an example on how to calculate an individual's average-based chart (subgroup size n = 1). In Table 6-6, data are listed for total chlorine at one sampling site in the order that they were measured. The average of the data is calculated.

Table 6-6 Calculation of an average-based Shewhart control chart with subgroup size n = 1

Date	Total Chlorine, mg/L	Moving Range
5/8/2015	2.90	—
5/9/2015	3.10	0.20
5/10/2015	3.10	0.00
5/11/2015	3.10	0.00
5/12/2015	3.30	0.20
5/13/2015	3.40	0.10
5/14/2015	3.10	0.30
5/15/2015	3.20	0.10
5/16/2015	3.00	0.20
5/17/2015	3.10	0.10
5/18/2015	3.40	0.30
5/19/2015	3.00	0.40
Average	3.14	0.17

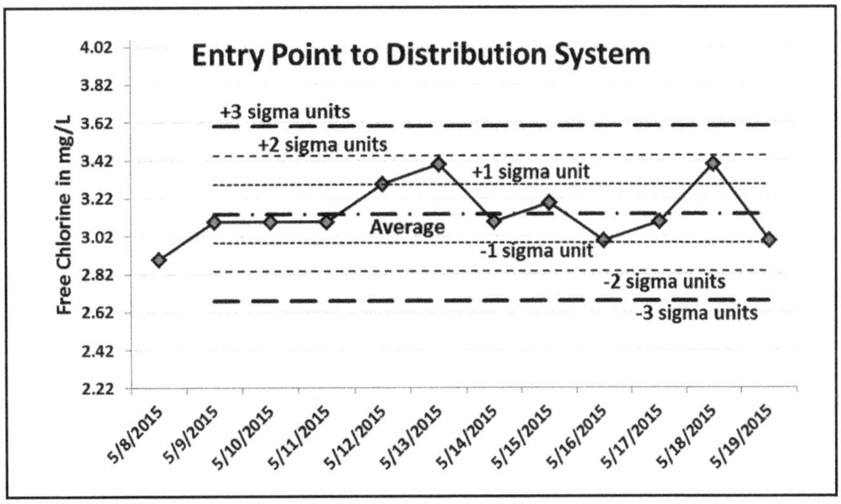

Figure 6-12 Shewhart control chart of Table 6-6 data

Next, the *moving range* is calculated by subtracting the value of one result from the previous result. The absolute value of the difference is used so that there are no negative numbers. Finally, the average of the moving range is calculated. The moving range provides an understanding of the variation from one measurement to the next.

The unit of variation, called the *sigma unit,* is based on the average moving range in the dataset. The sigma unit is calculated by dividing the average moving range by a special constant based on data subgroup size. The constants have been calculated based on the structure of data distributions, but like many concepts in statistics, the constants are conveniently listed in tables for use in constructing control charts (Wheeler and Chambers 1992). For the case of a dataset with a subgroup of size n = 1, the constant is 1.128. In the example of Table 6-6, the sigma unit (unit of variation) of the dataset is 0.17 divided by 1.128, which equals 0.15.

Now there is enough information to construct the "guidelines" behind the time series graph of total chlorine measurements. Figure 6-12 shows the Shewhart control chart for the data in Table 6-6. The calculations for the guidelines are as follows:

- Average = 3.14
- Sigma unit = 0.17/1.128 = 0.15
- –1 sigma unit = 3.14 – (1 × 0.15) = 2.99
- –2 sigma units = 3.14 – (2 × 0.15) = 2.84
- –3 sigma units = 3.14 – (3 × 0.15) = 2.68 = LCL
- +1 sigma unit = 3.14 + (1 × 0.15) = 3.29
- +2 sigma units = 3.14 + (2 × 0.15) = 3.45
- +3 sigma units = 3.14 + (3 × 0.15) = 3.60 = UCL

The area between the upper and lower control limits can be thought of as the expected range of values of the parameter if the system continues operating under the same conditions. If the measurements fall outside the expected range or if they form certain patterns that are highly improbable to occur, it is an indication that the conditions in the system are now different than before. The example in Figure 6-12 happens to be a system where the conditions have little variability. The data exhibit no improbable patterns. The graph is inspected for the following improbable trends to determine if the system conditions have changed:

- Data fall outside the 3 sigma unit lines.
- At least two out of three successive values fall on the same side of the average and are 2 sigma units or greater away from the average.
- At least four out of five successive values fall on the same side of the average and are 1 sigma unit or greater away from the average.
- Eight or more successive points fall on the same side of the average line.

These charts are easy to use, but they are also easy to misuse. The reader is referred to books on Shewhart control charts for a better understanding of individual charts and other applications of the control chart technique (Wheeler and Chambers 1992).

If a chart shows that an improbable trend has occurred or is occurring, the reasons that this occurred must quickly be determined by reviewing water system operation during the time period of the data trend. It can be anything from a faulty measurement sensor to a major event like a water main break. But every improvement in response to these data trends can add up to overall better water quality and control in the water system.

Shewhart control charts can also be used for water system process control and improvement, just as they are used in industry. As an example, a goal of maintaining an average pH of 7.4 has been set in a hypothetical water system. In addition, the pH is expected to vary, but it is desired to keep the variance between 7.2 and 7.6. The average and expected range can be plotted on a graph, where 7.6 will be equal to the UCL and 7.2 will be equal to the LCL. The actual pH data over time from the sampling site on top of these guidelines can also be plotted on the same graph. If the data exhibit any of the improbable trends listed above, changes must be made to the system operations until the data stay within the ideal expected variation. This method of process control and improvement is an iterative technique of changing operations, monitoring results, and then comparing results to the ideal statistics.

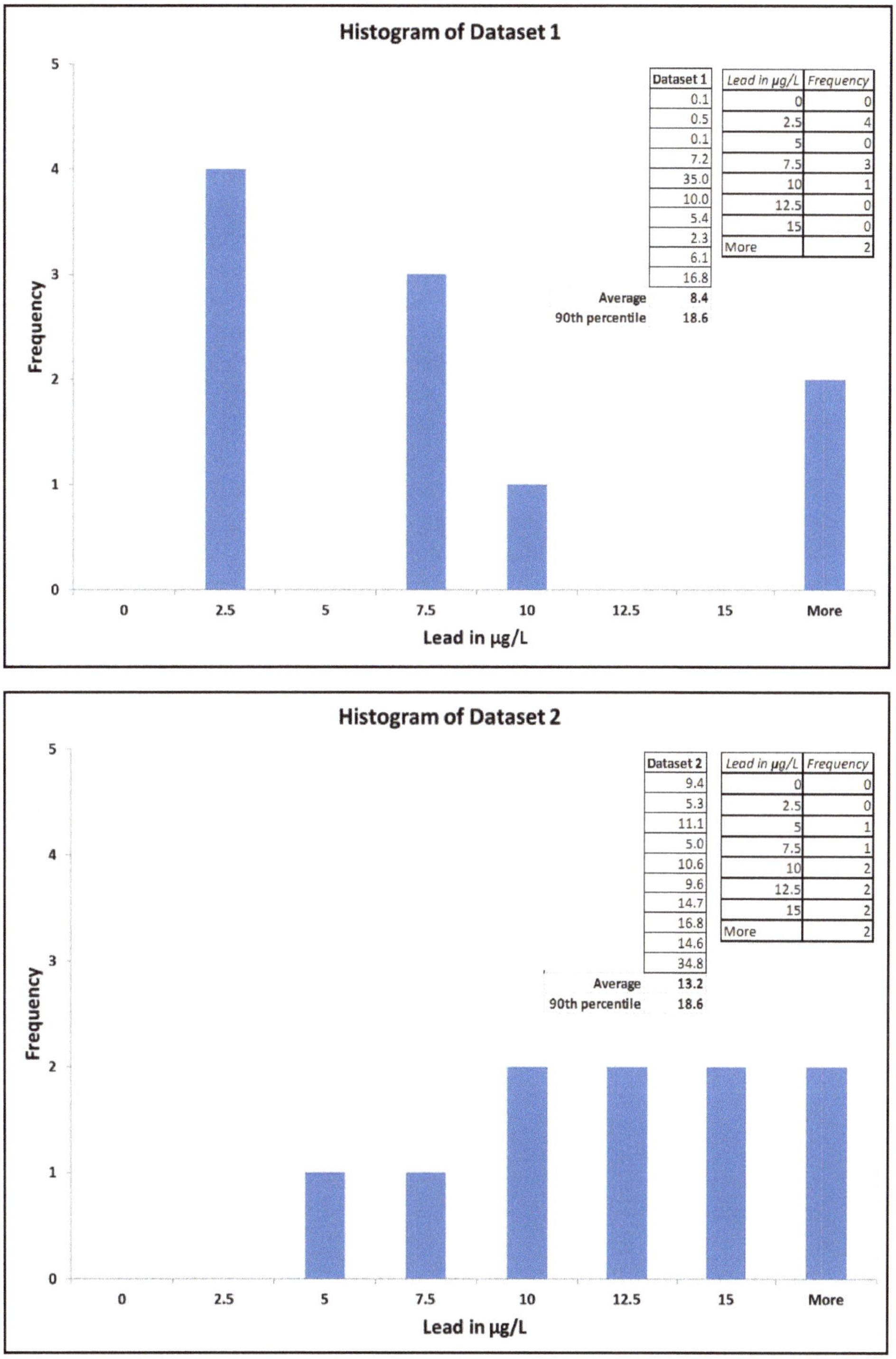

Figure 6-13 Histogram examples of two datasets of lead concentration data

Other Data Analytical Techniques

Other data analytical techniques can be helpful in analyzing water quality monitoring data and they should be used when applicable.

For example, as lead and copper release studies become more comprehensive with more water quality parameters measured, parameter co-trending techniques are being used. One such technique is Spearman's rank correlation (Wikipedia 2016). Using this technique, an assessment is made as to whether two parameters are increasing at the same time, decreasing at the same time, moving in opposite directions at the same time, or experiencing unrelated trends. When common trends are identified using this technique, causation is not implied because other factors could be influencing both parameters equally. Nevertheless, it is of interest to see which water quality parameters trend with lead or copper concentration released to the water. Such trending analysis will not be covered further in this manual but can be found in references on data analytical techniques.

As another example, it is informative to see the frequency that specific ranges of results occur in a set of data. Graphs of possible result ranges and the frequency with which they occur are called "histograms" (Box et al. 1978). Figure 6-13 shows lead concentration data from Lead and Copper Rule compliance sampling in two water systems. Both sets of data have the same 90th percentile lead concentration of 18.6 µg/L. However, Dataset 1 has only two sites out of ten where the lead concentration is over 10 µg/L, an arbitrary boundary in this example defining high lead sites. Dataset 2 has six sites out of ten where the lead concentration is over 10 µg/L. It is, therefore, informative to see, by use of histograms, that System 1 has elevated lead at specific buildings, about 20 percent of all buildings. System 2 has elevated lead as a more systemic problem at about 60 percent of all buildings. The approach to controlling lead in each system may be different based on this insight.

SUMMARY

The development of a monitoring plan to verify the effectiveness of corrosion control treatment and other metal release remediation includes three tools:

1. A flexible and comprehensive monitoring strategy. One such strategy groups water quality parameters into the three main categories of factors that can release metal from surfaces to water, as discussed in chapter 2.
2. Sampling sites and devices that are representative of water quality found in the distribution system, where it is important to use a combined strategy of selecting sites from sites that define the extremes of water quality in a system, individual problematic buildings connected to the distribution system, and sites that can establish a geographic grid of information.
3. A data analytical technique, such as Shewhart control charts, that can quickly make sense of the monitoring data and guide decisions for better control of metal release.

REFERENCES

ASTM D2688-83 Method B. 1983a. *Standard Test Methods for the Corrosivity of Water in the Absence of Heat Transfer (Weight Loss Protocol).* Philadelphia, Pa.: ASTM International.

ASTM D2688-83 Method C. 1983b. *Standard Test Methods for the Corrosivity of Water in the Absence of Heat Transfer (Machined Nipple Test).* Philadelphia, Pa.: ASTM International.

ASTM G1-81. 1981. *Recommended Practice for Preparing, Cleaning and Evaluating Corrosion Test Specimens.* Philadelphia, Pa.: ASTM International.

ASTM G46-76. 1976. *Recommended Practice for the Examination and Evaluation of Pitting Type Corrosion.* Philadelphia, Pa.: ASTM International.

AwwaRF and TZW (Awwa Research Foundation and DVGW-Technologiezentrum Wasser). 1996. *Internal Corrosion of Water Distribution Systems*, 2nd ed. Denver, Colo.: AwwaRF.

Box, G.E.P., W.G. Hunter, and J.S. Hunter. 1978. *Statistics for Experimenters*. New York: John Wiley and Sons.

Cantor, A.F. 2009. *Water Distribution System Monitoring: A Practical Approach for Evaluating Drinking Water Quality*. Boca Raton, Fla.: CRC Press.

Cantor, A.F. 2017. Optimization of Phosphorus-Based Corrosion Control Chemicals and Flushing for Lead and Copper Control. Denver, Colo.: Water Research Foundation, forthcoming.

Cantor, A.F., D. Denig-Chakroff, R.R. Vela, M.G. Oleinik, and D.L. Lynch. 2000. Use of polyphosphate in corrosion control. *Journal AWWA*, 92(2):95.

Cantor, A.F., J.K. Park, and P. Vaiyavatjamai. 2003. Effect of chlorine on corrosion in drinking water systems. *Journal AWWA*, 95(5):112.

Cantor, A.F., E. Kiefer, J.K. Little, A. Degnan, A. Jacque, J.B. Maynard, D. Mast, and J. Cantor. 2012. *Distribution System Water Quality Control Demonstration*. Denver, Colo.: Water Research Foundation.

Cornwell, D., R. Brown, and N. McTigue. 2015. Controlling Lead and Copper Rule water quality parameters. *Journal AWWA*, 107:2:E86.

EES (Economic and Engineering Services, Inc.). 1990. *Lead Control Strategies*. Denver, Colo.: AwwaRF.

Parks, J., M. Edwards, and A. Atassi. 2014. *Non-Intrusive Methodology for Assessing Lead and Copper Corrosion*. Denver, Colo.: Water Research Foundation.

Reiber, S.H., J.F. Ferguson, and M.M. Benjamin. 1988. An improved method for corrosion rate determination by weight loss. *Journal AWWA*, 79(2):71.

Temkar, P. 1988. *Development of a Pipe Loop System for Determining Effectiveness of Corrosion Control Chemicals in Potable Water Systems*. USA-CERL Technical Report N-88/12. Champagne, Ill.: US Army Corps of Engineers Research Laboratory.

Temkar, P., R.J. Schloze, S.W. Maloney, and C.H. Neff. 1987. Pipe loop system for evaluating effects of water quality control chemicals in water distribution systems. In *Proceedings of the AWWA Water Quality Technology Conference*. Denver, Colo.: AWWA.

Wheeler, D.J., and D.S. Chambers. 1992. *Understanding Statistical Process Control*, 2nd ed. Knoxville, Tenn.: SPC Press.

Wikipedia. 2016. Spearman's rank correlation coefficient. Available at: https://en.wikipedia.org/wiki/Spearman%27s_rank_correlation_coefficient

Williams, S.M., R.G. Ainsworth, and A.F. Elvidge. 1984. *A Method of Assessing the Corrosivity of Waters Towards Iron*. Swindon, U.K.: Water Research Centre.

This page intentionally blank.

Index

NOTE: *f.* indicates figure; *t.* indicates table

D

M

N

O

P

Q

R

S

T

U

W

X

Z

AWWA Manuals

M1, *Water Rates, Fees, and Charges,* #30001
M2, *Instrumentation and Control,* #30002
M3, *Safety Management for Water Utilities,* #30003
M4, *Water Fluoridation Principles and Practices,* #30004
M5, *Water Utility Management,* #30005
M6, *Water Meters—Selection, Installation, Testing, and Maintenance,* #30006
M7, *Problem Organisms in Water: Identification and Treatment,* #30007
M9, *Concrete Pressure Pipe,* #30009
M11, *Steel Pipe—Design and Installation,* #30011
M12, *Simplified Procedures for Water Examination,* #30012
M14, *Backflow Prevention and Cross-Connection Control: Recommended Practices,* #30014
M17, *Installation, Field Testing, and Maintenance of Fire Hydrants,* #30017
M19, *Emergency Planning for Water Utilities,* #30019
M20, *Water Chlorination/Chloramination Practices and Principles,* #30020
M21, *Groundwater,* #30021
M22, *Sizing Water Service Lines and Meters,* #30022
M23, *PVC Pipe—Design and Installation,* #30023
M24, *Planning for the Distribution of Reclaimed Water,* #30024
M25, *Flexible-Membrane Covers and Linings for Potable-Water Reservoirs,* #30025
M27, *External Corrosion Control for Infrastructure Sustainability,* #30027
M28, *Rehabilitation of Water Mains,* #30028
M29, *Water Utility Capital Financing,* #30029
M30, *Precoat Filtration,* #30030
M31, *Distribution System Requirements for Fire Protection,* #30031
M32, *Computer Modeling of Water Distribution Systems,* #30032
M33, *Flowmeters in Water Supply,* #30033
M36, *Water Audits and Loss Control Programs,* #30036
M37, *Operational Control of Coagulation and Filtration Processes,* #30037
M38, *Electrodialysis and Electrodialysis Reversal,* #30038
M41, *Ductile-Iron Pipe and Fittings,* #30041
M42, *Steel Water-Storage Tanks,* #30042
M44, *Distribution Valves: Selection, Installation, Field Testing, and Maintenance,* #30044
M45, *Fiberglass Pipe Design,* #30045
M46, *Reverse Osmosis and Nanofiltration,* #30046
M47, *Capital Project Delivery,* #30047
M48, *Waterborne Pathogens,* #30048
M49, *Quarter-Turn Valves: Torque, Head Loss, and Cavitation Analysis,* #30049
M50, *Water Resources Planning,* #30050
M51, *Air-Release, Air/Vacuum, and Combination Air Valves,* #30051
M52, *Water Conservation Programs—A Planning Manual,* #30052
M53, *Microfiltration and Ultrafiltration Membranes for Drinking Water,* #30053
M54, *Developing Rates for Small Systems,* #30054
M55, *PE Pipe—Design and Installation,* #30055
M56, *Nitrification Prevention and Control in Drinking Water,* #30056
M57, *Algae: Source to Treatment,* #30057
M58, *Internal Corrosion Control in Water Distribution Systems,* #30058
M60, *Drought Preparedness and Response,* #30060
M61, *Desalination of Seawater,* #30061
M63, *Aquifer Storage and Recovery,* #30063
M65, *On-Site Generation of Hypochlorite,* #30065
M66, *Cylinder and Vane Actuators and Controls—Design and Installation,* #30066

www.ingramcontent.com/pod-product-compliance
Lightning Source LLC
LaVergne TN
LVHW070107021026
849160LV00026B/1698

9781625762023